Color Atlas of

MOUNTAIN MEDICINE

1 Cover illustration: a severely injured mountaineer receives on-the-spot treatment from a doctor and his medical support team, before transportation to hospital. *Courtesy of Bachelet/publications filipacchi.*

Color Atlas of MOUNTAIN MEDICINE

Editors

Frédéric Dubas
MD, FMH
General Surgeon
ISMM-General Secretary
CH-1950 Sion
Switzerland

Jacques Vallotton
MD
Department of Orthopedic Surgery
University Hospital
CH-1011 Lausanne
Switzerland

in collaboration with
The International Society for Mountain Medicine
(ISMM)

Mosby Year Book

St. Louis Baltimore Boston Chicago London Philadelphia Sydney Toronto

Mosby Year Book
Dedicated to Publishing Excellence

Mosby–Year Book, Inc.
11830 Westline Drive
St. Louis, MO 63146

Published in 1991 with rights in the USA, Canada and Puerto Rico by Mosby–Year Book, Inc.

ISBN 0 8151 8953 2

English edition first published in 1991 by Wolfe Publishing Ltd, 2–16 Torrington Place, London WC1E 7LT, UK.

Library of Congress Cataloging-in-Publication Data

A Colour atlas of mountain medicine / edited by Jacques Vallotton.
p. cm.
Includes index.
ISBN 0–8151–8953–2
1. Mountaineering--Health aspects--Atlases. 2. Cold--Physiological effect--Atlases. 3. Altitude, Influence of--Atlases.
I. Vallotton, Jacques.
[DNLM: 1. Mountaineering--atlases. 2. Sports Medicine--atlases.
QT 17 C719]
RC1220.M6C65 1991
616.9'8--dc20
DNLM/DLC
for Library of Congress 91-19681
CIP

Contents

Foreword 7
John B. West

Acknowledgments 8

List of Contributors 9

Introduction 11

Historical aspects of mountain medicine 13

Early scientific expeditions to high altitude 13
John B. West

Mountain rescue today 17

The helicopter pioneers of modern mountain rescue 17
Bruno Bagnoud
Mountain rescue: modern strategies 18
Bruno Durrer
Technical aspects of mountain rescue 22
Rémy Henzelin
Medical assistance in mountain rescue 34
Frédéric Dubas

Part I Climatic conditions and associated pathologies

Altitude 43

Definition and parameters 43
Paolo Cerretelli
Performance and metabolism at high altitude 46
Paolo Cerretelli
Training for high-altitude climbing 48
Oswald Oelz
Nutrition at altitude 49
Paolo Cerretelli
Sleep at altitude 50
Paolo Cerretelli
Acute mountain sickness 51
Oswald Oelz
Acute mountain sickness: risk factors 54
Jean-Paul Richalet
Localized or peripheral edema 57
Oswald Oelz
Polycythemia and thromboembolic episodes 57
Oswald Oelz
High-altitude pulmonary edema 58
Oswald Oelz
High-altitude cerebral edema 60
Charles Clarke
High-altitude retinopathy 62
Charles Clarke
Mental and neurological disturbances at high altitude 64
Zdzislaw Ryn

Solar radiation 67

The effects of solar radiation 67
Peter Lomax, Richard Thinney and Bartly J. Mondino
Eye protection at high altitude 71
Dominique Petetin

Cold 73

Frostbite and mountaineering 73
Bernard Marsigny and Jacques Foray
Frostbite 78
William J. Mills, Jr
Hypothermia 92
Bruce C. Paton
Hypothermia in an avalanche, a case report 96
Adam P. Fischer, Frank Stumpe and Jacques Vallotton
Hypothermia in a crevasse, a case report 98
Frédéric Dubas and Jürg Pfenninger

Ice and snow 99

Personal experience of an avalanche 99
Georgy Vianin
Avalanches and their mechanisms 100
André Roch

Avalanches: prevention and rescue **104**
Frédéric Dubas, Rémy Henzelin and Jacques Michelet
Rescue in crevasses **112**
Frédéric Dubas, Rémy Henzelin and Jacques Michelet

Lightning **117**
Olivier Baptiste, Alain Girer and Jacques Foray

Part II Traumatology and mountain sports

Hiking and trekking **123**

Injuries related to hiking and trekking **123**
Albino Lanzetta and Remo Seratoni
Muscular lesions **128**
Alain Rostan and Charles Gobelet

Downhill skiing **135**

Downhill skiing accidents: prevention and statistics **135**
José M. Figueras, Aleix Vidal and Eulalia Escola
The management of skiing accidents **141**
Henri Bèzes, Pierre Massart and Catherine Guyot
Injuries to the ligaments of the knee **152**
Pierre-François Leyvraz
Capsuloligamentary injuries to the metacarpophalangeal joint of the thumb ('skier's thumb') **155**
José Cantero and Jacques Vallotton
Monoskiing and snow surfing: new risks **160**
Marc-Hervé Binet

Mountaineering **165**

Forces on the human body during falls on the rope and their consequences **165**
Helmut Mägdefrau
Head injuries **171**
Jacques de Preux
Spinal injuries **173**
Jacques de Preux
Injuries in sport climbing **176**
Conxita Leal and Anton Rañé
Training procedures and basic principles to avoid the risk of injury in modern free climbing **182**
Lorenz Radlinger
Specific aspects of cave rescue **187**
Olivier Moeschler
Proposal for a modular first aid kit for climbers, mountain guides and climbing doctors **195**
Urs Wiget
Paragliding **200**
Stéphane Oggier

Future prospects of mountain medicine **206**
Frédéric Dubas and Jacques Vallotton

References **211**

Index **219**

Foreword

It is a pleasure and privilege to contribute a foreword to *Color Atlas of Mountain Medicine.* More and more people are becoming aware of the recreational opportunities in mountain areas as the number of trekkers, skiers and climbers continues to increase. The ease of international travel means that mountainous areas are available to thousands of people who previously did not consider them suitable for vacations. Indeed, travel to exotic, remote regions of the world is becoming astonishingly easy. Whereas only the most adventuresome would travel to the remoter parts of Asia and South America a few years ago, such places are now within relatively easy reach. The pace of travel, too, has quickened dramatically. Europeans can find themselves in Lhasa or Cuzco within a day or two of leaving home. The remote regions of the world have never been so accessible.

The easy access of mountain regions means that accidents and emergencies are disturbingly common. Increasingly, people who have not previously visited high areas find themselves surprised by rapid changes of weather conditions in an environment which is notoriously treacherous. Fortunately, modern methods of rescue have become much more sophisticated and this book provides a timely and comprehensive account of the state of the art. In many respects the mountain rescue procedures developed in Switzerland have served as a model for other parts of the world.

On a final note, travelers in general, and especially those to mountainous areas, have often contributed to the international contacts which help people to live together in a divided world. It has frequently been said that though mountains divide countries, they unite people. In an increasingly complex and fragmented world, the camaraderie of travelers to mountains can be a catalyst for better international relationships.

I wish this handsome book well, and hope that it will help to make mountain places safer for the increasingly large number who enjoy them.

John B. West
President, ISMM
La Jolla, California

Acknowledgments

It was in Chamonix, after an adventurous run down the Vallée Blanche in the spring of 1984, that Dr Augustin Besson gave Dr Jacques Vallotton the idea of preparing a color atlas on mountain accidents. Little by little the project took shape; then, after meeting Dr Frédéric Dubas in 1986, the book acquired its present form. Their friendship and tireless efforts finally got this book to the printers.

They would particularly like to thank:

The International Society for Mountain Medicine (ISMM) for its cordial hospitality and for giving this book an international dimension.

Mrs Ariane Berti for typing (and retyping) the manuscript.

Professor James Freeman, Drs Marc Bernath, Michael Dusmet, Marianne Lorenz, Athanassios Salessiotis and John Wellinger, Mr Roger Anthoine, Mrs Elizabeth Dubas and Martha Haldimann for their help with translations.

Véronique Michel, Renaud de Watteville, Frédéric Mauclet, Marie-Hélène Schmidt, Evelyne Wirth, Rémy Wenger, the Swiss Air Rescue Guard (REGA), and all the others for their excellent illustrations for this book.

Maya and France, their wives, for their encouragement and patience throughout this project.

List of contributors

Bruno Bagnoud, Director, 'Air-Glaciers', SA, Regional Airport, CH-1950 Sion, **Switzerland**.

Dr Olivier Baptiste, Hôpital de Chamonix, F-74403 Chamonix–Mt. Blanc, **France**.

Professor Henri Bèzes, Department of Orthopedic Surgery and Traumatology, University Hospital, F-38130 Echirolles, **France**.

Dr Marc-Hervé Binet, Medical Center, Le Pas du Lac, F-74110 Avoriaz, **France**.

Dr José Cantero, Clinique et permanence de Longeraie, Av. de la Gare 9, CH-1004 Lausanne, **Switzerland**.

Professor Paolo Cerretelli, Department of Physiology, former President International Society for Mountain Medicine, University Medical Center, 5, Rue Michel-Servet, CH-1205 Geneva, **Switzerland**.

Dr Charles Clarke, President Medical Commission of the International Union of Alpinists' Associations, Department of Neurological Sciences, St Bartholomew's Hospital, 38, Little Britain, London EC1A 7BE, **United Kingdom**.

Dr Frédéric Dubas, General Surgeon, Mountain Rescue Doctor, General Secretary International Society for Mountain Medicine, 4, Av. de la Gare, CH-1950 Sion, **Switzerland**.

Dr Bruno Durrer, General Practitioner, Mountain Guide, Mountain Rescue Doctor, Aerztehuus, CH-3822 Lauterbrunnen, **Switzerland**.

Dr Eulalia Escola, Department of Orthopedic Surgery and Traumatology, Hospital de la Cruz Roja, E-08025 Barcelona, **Spain**.

Professor José M. Figueras, Department of Orthopedic Surgery and Traumatology, Hospital de la Cruz Roja, E-08025 Barcelona, **Spain**.

Dr Adam P. Fischer, Department of Cardiovascular Surgery, University Hospital, CH-1011 Lausanne, **Switzerland**.

Dr Jacques Foray, Hôpital de Chamonix, F-74403 Chamonix–Mt. Blanc, **France**.

Dr Alain Girer, Hôpital de Chamonix, F-74403 Chamonix–Mt. Blanc, **France**.

Dr Charles Gobelet, Center for Sports Medicine, Department of Physical Medicine and Rehabilitation, Hôpital de Gravelone, CH-1950 Sion, **Switzerland**.

Dr Catherine Guyot, Department of Orthopedic Surgery and Traumatology, University Hospital, F-38130 Echirolles, **France**.

Rémy Henzelin, Retired test pilot of the Swiss Army, 'Air-Glaciers', SA, Regional Airport, CH-1950 Sion, **Switzerland**.

Dr Albino Lanzetta, Clinica Orthopedica Università degli Studi di Milano, P. 77a C. Ferrari 1, 20122 Milano, **Italy**.

Dr Conxita Leal Tort, Medicina de l'esport, Muntaner 231, 2'2', E-08021 Barcelona, **Spain**.

Dr Pierre-François Leyvraz, Orthopedic Surgeon, Department of Orthopedic Surgery and Traumatology, University Hospital, CH-1011 Lausanne, **Switzerland**.

Professor Peter Lomax, University College of Los Angeles, School of Medicine, Los Angeles, California 90024, **United States of America**.

Helmut Mägdefrau, Stefanienstrasse 6, D-8024 Deisenhoffen, **Germany**.

Dr Bernard Marsigny, Hôpital de Chamonix, F-74403 Chamonix–Mt. Blanc, **France**.

Dr Pierre Massart, Department of Orthopedic Surgery and Traumatology, University Hospital, F-38130 Echirolles, **France**.

Jacques Michelet, Mountain Guide, Chief of the Rescue and Research Service of the Cantonal Police of the Canton of Valais, Av. de France, CH-1951 Sion, **Switzerland**.

Professor William J. Mills, Jr, Director, The Center for High Latitude Health Research, University of Alaska, 3211 Providence Drive, Anchorage, Alaska 99508, **United States of America**.

Dr Olivier Moeschler, Air Rescue Doctor, Department of Anaesthesiology, University Hospital, CH-1011 Lausanne, **Switzerland**.

Dr Bartly J. Mondino, Professor of Ophthalmology, Jules Stein Eye Institute, University of California, Los Angeles, California 90024, **United States of America**.

Professor Oswald Oelz, Department of Internal Medicine, University Hospital, Rämistrasse, 100, CH-8091 Zurich, **Switzerland**.

Dr Stéphane Oggier, General Practitioner, Mountain Guide, Ch. de la Sitterie, 7, CH-1950 Sion, **Switzerland**.

Professor Bruce C. Paton, Department of Cardiothoracic Surgery, 950, East Harvard Avenue, Suite 680, Denver, Colorado 80210, **United States of America.**

Dominique Petetin, Optician, Schmutz Opticians, 20 Petit-Chêne, CH-1002 Lausanne, **Switzerland.**

Dr Jürg Pfenninger, Children's Intensive Care Unit, University Hospital, CH-3010 Berne, **Switzerland.**

Dr Jacques de Preux, FMH, Neurosurgeon, Department of Neurosurgery, Regional Hospital, CH-1950 Sion, **Switzerland.**

Lorenz Radlinger, Biologist, Birkenweg, 15, CH-3072 Ostermundigen, **Switzerland.**

Dr Anton Rañé, Medecina de l'esport, Muntaner 231, 2'2', E-08021 Barcelona, **Spain.**

Professor Jean-Paul Richalet, Association de Recherche pour la Physiologie de l'Environnement, Laboratoire de Physiologie, UFR Médecine, 74, Rue Marcel-Cachin, F-93012 Bobigny, **France.**

André Roch, Engineer, Avalanche Specialist, 3, Ch. Naville, CH-1231 Conches-Geneva, **Switzerland.**

Dr Alain Rostan, 19 Rue de Collège, 1227 Carouge, Geneva, **Switzerland.**

Professor Zdzislaw Ryn, Department of Social Pathology, Medical Academy, ul. Skarbowa 4, 31-121 Krakow, **Poland.**

Dr Remo Seratoni, Clinica Orthopedica, Università degli Studi di Milano, P. 77a C. Ferrari 1, 20122 Milano, **Italy.**

Dr Frank Stumpe, Department of Cardiovascular Surgery, University Hospital, CH-1011 Lausanne, **Switzerland.**

Dr Richard Thinney, Jules Stein Institute, University of California, Los Angeles, California 90024, **United States of America.**

Dr Jacques Vallotton, Department of Orthopedic Surgery and Traumatology, University Hospital, CH-1011 Lausanne, **Switzerland.**

Georgy Vianin, Professional Mountain Guide and Ski Instructor, CH-3961 Zinal, **Switzerland.**

Dr Aleix Vidal, Department of Orthopedic Surgery and Traumatology, Hospital de la Cruz Roja, E-08025 Barcelona, **Spain.**

Professor John B. West, Department of Physiology, School of Medicine, University College of San Diego, La Jolla, California 92093, **United States of America.**

Dr Urs Wiget, General Practitioner, Mountain Rescue Doctor, President Medical Commission (CISA–IKAR), President Medical Mountain Rescue Team (GRIMM), CH-3961 Vissoie, **Switzerland.**

Introduction

The prevention of accidents and life-threatening situations is the hallmark of modern mountain medicine. The arrival of new 'high-risk' mountain sports, as well as the increasing popularity of traditional recreations, such as skiing and mountain climbing, make this preventive aspect even more important.

In these pages you will find not only chapters on the medical problems brought about by altitude and climate, but information on first aid, the treatment of specific injuries, mountain rescue, and a synthesis of the latest research on altitude physiology.

The mountain has a continuing fascination for men and women and many new sports have developed, in addition to the more traditional mountain climbing and skiing. Whatever challenge it offers, the mountain dictates its own law. Any lack of understanding or misjudgement of its nature, or human failure, can rapidly lead to a catastrophe. As the number of accidents increases, many courageous men and women risk their own lives to save the wounded or ill. The correct link between understanding and prevention can help to minimize these risks.

Mountain medicine is still in many ways an empirical field, but the many expeditions to the Himalayas and the Andes have contributed greatly to our understanding of physiological adaptation to altitude. This book brings together the most up-to-date knowledge on this subject. The contributors are all specialists in their own fields, and most are associated with the International Society for Mountain Medicine (ISMM).

Dedication

To our wives, France and Maya,
and our children, Céline, Julie, Laurent, Loïse, Olivier and Stéphanie:
future lovers of safe mountaineering.

Historical aspects of mountain medicine

Early scientific expeditions to high altitude

John B. West

One of the earliest scientific expeditions to high altitude was the ascent of Mont Blanc (4,807 m) in 1787 by Professor de Saussure, a Genevan physicist (Freshfield, 1920). In **2**, his guides are shown carrying scientific equipment for the measurement of temperature, snow conditions, etc., on the summit. De Saussure complained bitterly of the effects of high altitude: 'the kind of fatigue which results from the rarity of the air is absolutely unconquerable; when it is at its height, the most terrible danger would not make you take a single step further'. This was the second ascent of the highest mountain in the European Alps, the first having been made by Paccard and Balmat the previous year.

A landmark in the history of high-altitude medicine was the construction of Capanna Regina Margherita in 1893 by Angelo Mosso, Professor of Physiology at the University of Turin (**3**). Funds for the building were provided by Queen Margherita of Italy, an ardent mountaineer herself. The laboratory was placed on the Punta Gnifetti on Monte Rosa at an altitude of 4,559 m, and is still in use today.

Apparently, the first physiological expedition to high altitude was that organized by the German physiologist, Nathan Zuntz, to Mt. Tenerife in the Canary Islands in 1910. The highest station was the Alta Vista Hut at about 3,350 m, and the results suggested that hypoxia was the most important factor in high-altitude disability, as predicted by Paul Bert in 1878.

A year earlier, in 1909, the Italian aristocrat, the Duke of the Abruzzi, had reached the remarkable altitude of 7,500 m in the Karakoram Mountains. Although this was not primarily a scientific enterprise, one of the reasons given for the expedition was to determine 'the greatest height to which man may attain in climbing' (Filippi, 1912), and the climb prompted the British physiologists

2

2 A contemporary print of Professor de Saussure and his guides, during their ascent of Mont Blanc in 1787. (There is some question about whether this is actually Mont Blanc.)

3

3 Capanna Regina Margherita on the Punta Gnifetti on Monte Rosa, at an altitude of 4,559 m.

4

4 C.G. Douglas, J.S. Haldane, Y. Henderson and E.C. Schneider carrying out measurements during the 1911 Anglo-American Pike's Peak Expedition, in Colorado.

5

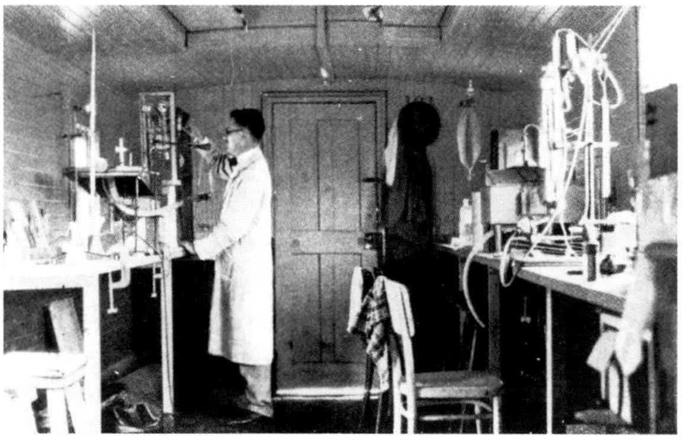

5 Laboratory inside a railway baggage car used by Barcroft and his colleagues in Cerro de Pasco, 4,330m.

6

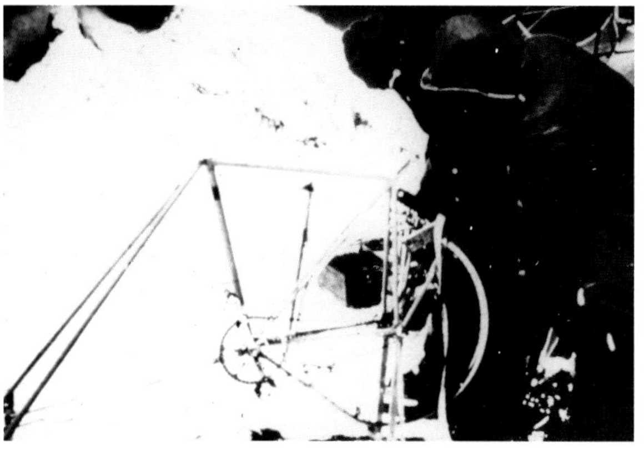

6 Bicycle ergometer being assembled on the Makalu Col (7,440m) in 1961. These experiments gave the highest measurements of maximal oxygen uptake to date.

Douglas and Haldane to conclude that oxygen secretion by the lung must have occurred (Douglas *et al.*, 1913). This hypothesis was tested on the Anglo-American Pike's Peak Expedition in Colorado in 1911 (**4**), and it was concluded that oxygen secretion did, in fact, occur at high altitude. However, this conclusion was challenged by another English physiologist, Joseph Barcroft, who led an important expedition to Cerro de Pasco in Peru in 1921–1922 (Barcroft *et al.*, 1923) (**5**). They found no evidence for oxygen secretion, but were enormously impressed by the residents' capacity for physical work at an altitude of 4,330m.

At about this time the Scottish physiologist, Alexander Kellas, carried out an analysis of the physiological problems of climbing Mt. Everest, and concluded that this would probably be possible without supplementary oxygen. Kellas' work has been almost completely overlooked (West, 1987); he died during the first reconnaissance expedition to Mt. Everest in 1921, just as the expedition had its first view of the mountain. Three years later, Norton reached an altitude of about 8,580m on Mt. Everest without supplementary oxygen, and it is remarkable that the last 300m were not conquered without oxygen until Messner and Habeler's classic ascent in 1978, some 54 years later.

An important international high-altitude expedition took place in 1935 under the scientific leadership of D. Bruce Dill. The site was Aucanquilcha in the Chilean Andes, where there is a sulfur mine which was said to be at an altitude of 5,800m. The primary purpose of this expedition was to study the changes in blood chemistry that occur at high altitude (Dill *et al.*, 1937). The Aucanquilchan miners have often been referred to as the highest inhabitants of the world. The expedition reported that they lived at an altitude of 5,340m and climbed 500m each day to work in the mine. However, a recent visit to the mine resulted in the surprising finding that a small group of miners live indefinitely at the mine itself, which is actually at an altitude of 5,950m (West, 1986). Only a small number of miners are able to tolerate this altitude; nevertheless, this makes them the highest inhabitants of the world.

In 1952, the British physiologist L.G.C.E. Pugh carried out field measurements on Cho Oyu near Mt. Everest, in preparation for what was to be the first successful ascent in 1953. An important factor in this expedition was the development of adequate oxygen equipment. Pugh also organized the Himalayan Scientific Mountaineering Expedition of 1960–1961, during which a group of eight physiologists wintered at an altitude of 5,800 m in a prefabricated hut, subsequently known as the 'Silver Hut'. In the following spring, the party attempted to climb Mt. Makalu (8,475 m) without supplementary oxygen, and though they failed in this, measurements of maximal oxygen consumption using a bicycle ergometer were made, up to an altitude of 7,440 m (Pugh *et al.*, 1964) (**6**).

In 1973, an Italian Mt. Everest expedition set up a physiological laboratory at the base camp, altitude 5,350 m, under the scientific leadership of Paolo Cerretelli. Measurements were obtained from climbers who had been above 8,000 m. One of the interesting observations was that breathing 100% oxygen at this altitude did not restore the maximal oxygen uptake to its pre-expedition sea level value. During the period 1968–1979, high-altitude studies were also carried out on Mt. Logan (5,334 m) in the Canadian Yukon, under the scientific leadership of C.S. Houston. Many contributions were made to our knowledge of mountain sickness, especially the high frequency of retinal haemorrhages.

The gathering of medical data at altitudes over 8,000 m was also one of the main objectives of the American Medical Research Expedition to Everest in 1981. Dr Christopher Pizzo obtained the first alveolar gas samples on the summit (**7**), and made the first direct measurement of barometric pressure. Both provided some surprises (West, 1985). His alveolar $P\text{CO}_2$ was only 1 kPa on the summit, indicating an astonishingly high level of ventilation. The barometric pressure was 33.7 kPa, far higher than that predicted from the standard altitude–pressure tables, which had previously been used by some physiologists.

Though not field expeditions, two simulated ascents of Mt. Everest have provided valuable data. Operation Everest I and II took place in 1946 and 1985, respectively, and both were planned by C.S. Houston. Among the important findings of Operation Everest II was the fact that myocardial contractility is well maintained up to altitudes near the summit of Mt. Everest, in spite of an arterial $P\text{O}_2$ of less than 4 kPa (Houston *et al.*, 1987).

In spite of this colorful history of scientific expeditions to high altitude, there remains much to be learned. Three topics that cry out for further study include the mechanism of muscle protein loss at high altitude, the acid–base status of climbers at extreme altitude, and the residual impairment of the central nervous system function after ascents to very high altitude. It is difficult to think of a better way of combining business with pleasure than a well-planned scientific expedition to high altitude.

7 Dr Christopher Pizzo taking alveolar gas samples on the summit of Everest, during the 1981 American Medical Research Expedition to Everest.

7

Mountain rescue today

The helicopter pioneers of modern mountain rescue

Bruno Bagnoud

The helicopter is the only means of transport that does not require any infrastructure: cars need roads, trains need rails, aircraft need runways, and boats can move only on water. Roads, railways and runways can be rendered useless by snow, floods, landslides or earthquakes, making the above modes of transport unsuitable for certain terrains. The helicopter's ability to land on and take off from a variety of terrains, on platforms measuring as little as two meters square (**8**), has made it an irreplaceable tool in mountain rescue.

In 1955, Jean Moine achieved a first when he landed his helicopter on the summit of Mont Blanc – a height of 4,807 m. This performance proved that modern helicopters had overcome the restrictions imposed by the rarefied air at high altitudes, and saw the beginning of a new era for transport in the mountains (until then, everything had been carried on men's backs, as mules seldom climb above 2,500 m). The first airborne mountain rescue took place in 1957 at the Vallot Refuge (4,362 m), with an *Alouette II* turbine helicopter, which opened the way to a new generation of machines.

In May 1952, Hermann Geiger developed a technique for landing on glaciers, using a Piper Super Cub (**9**), and went on to organize instruction courses. A contemporary, Fernand Martignoni, also contributed greatly to improvements in airborne rescue techniques.

The landing of an *Alouette III*, piloted by the author, on the extreme tip of the Matterhorn (4,476 m) in the summer of 1966, was undeniable proof of the effectiveness of this machine in mountain rescue.

8

8 Since the late 70s, over 90% of rescue operations in the Alps have been carried out using helicopters.

9

9 The first landing and take off on a glacier, performed by the Swiss pilot, Hermann Geiger, on the Kander glacier in 1952.

Mountain rescue: modern strategies

Bruno Durrer

A successful mountain rescue operation requires efficient communication, transport, evacuation and medical treatment, and favorable weather conditions. A long delay between the time of the accident and the raising of the alarm, together with a serious injury, has a negative effect on the outcome of the operation.

$$\text{Optimal rescue} = \frac{\text{Efficiency: 1 Communication, 2 Transport, 3 Evacuation, 4 Medical treatment}}{\text{Time lapse (accident – alarm)}} \times \frac{\text{Favorable meteorology}}{\text{Seriousness of injury}}$$

Communication

A rapid raising of the alarm saves time. Therefore, all mountain huts should have radiotelephones, and mountain guides and climbers should carry radios on all excursions. In Switzerland, relay stations throughout the Alps allow the alarm to be raised quickly. A central alarm station puts the nearest rescue helicopter in charge of the operation. If necessary, additional rescue specialists or equipment are directed to the site of the accident. Finally, the hospitals are given information about the number of victims and the type of injuries as soon as possible.

An efficient communication system is vital if delay is to be avoided between the time of the accident, the raising of the alarm and the treatment.

Transport

The helicopter, equipped with a winch, and preferably twin-engined, is the most efficient means of transport in the mountains. In Switzerland, a dense network of rescue stations allows helicopters to reach every possible accident site within 15 minutes. During the day, they can be airborne within five minutes, and at night, within 30 minutes.

Evacuation

Over 90% of all rescues in the high Alps (1,500/year) are performed using helicopters. Of these, 5% are combined rescues, e.g. the helicopter carries the rescuers below the cloud level, near the site of the accident. Only 5% of all mountain rescues are purely ground rescues, mainly due to bad visibility.

In 50% of all rescues in pathless areas, the helicopter is able to land; the remaining 50% are winch rescues. According to the type of accident (direct face, crevasse, avalanche, paragliding), different rescue specialists are brought to the site (**10** & **11**).

10

10 A helicopter performing a direct-face winch rescue. *Courtesy of Swiss Air Rescue.*

11

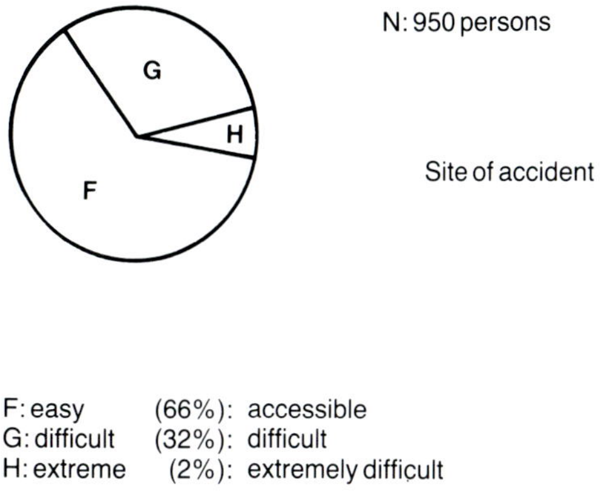

11 Topographical statistics of helicopter winch rescues.

Direct face rescues

Over 95% of all injured climbers in difficult sites are rescued by direct winch operation. On vertical faces, a winch extension measuring up to 70 or 80m is often employed. On overhanging rock faces, rescue specialists (mountain guides) are dropped as near as possible to the victim. The rescue is carried out using steel cable or static ropes, from the top. On north faces (e.g. the Eiger North Face (**12–14**)), the rescuers are, whenever possible, winched to a position from which they have access to the site of the accident. From here, they are able to climb to the injured and lower or pull them to a spot from which a winch rescue can be performed. The new direct face rescue techniques have now more or less taken the place of steel cable rescue operations on big faces.

Uninjured climbers on vertical faces are best evacuated without first lowering a rescuer. A radio is winched to the climbers, from which they receive instructions, and they then attach themselves to the winch cable. In this way, the rescue risks are minimized.

12

13

14

12–14 A difficult rescue on the north face of the Eiger.

Paragliding accidents

In the Alps, paragliding accidents have greatly increased over the last five years (*see also* pp. 200–205). The injured paragliders are often caught on rock or ice faces, suspended only by their parachute strings. In these cases, a normal winch rescue is not possible, because the downwash of the helicopter may provoke a further fall. To reduce the downwash, the winch is extended as much as 60 or 70 m. The rescuers are often dropped 50 m to the side of the injured person, and have to traverse the rock face to reach him. As soon as the parachute has been folded, a direct winch evacuation is possible.

Medical treatment

Optimal medical treatment starts at the site of the accident and is applied by a doctor, trained in anesthesiology and alpine techniques. The combination of a doctor and a mountain guide is ideal and often accelerates a rescue operation.

In recent years, the improved alarm system has led to the faster arrival of rescue teams at the site of the accident, allowing more severely injured patients to be treated more quickly in the mountains. The presence of an emergency doctor is now compulsory.

In direct face rescue, the rescuers are regularly faced with the dilemma of whether immediate evacuation or immediate treatment should have priority. Meteorological conditions, aspects of mountain safety and the type of injury all influence this decision. A severely injured climber should not be evacuated on his climbing harness. The least traumatic evacuation for the victim is in the rescue net or sac. However, maneuvering a patient into this net while on a difficult rock face often requires additional helpers.

During the evacuation the doctor performs perfusions, gives medication or even intubates the patient. 1.5% of all winch-rescued patients are intubated and ventilated during evacuation in the rescue net. After evacuation in the net, the helicopter lands and the patient is put on the stretcher inside. During transportation to the hospital, full monitoring is possible. In the Alps the hospitals can be reached within 15–30 minutes of take-off.

For the diagnosis and initial treatment of hypothermia, special thermometers and equipment for applying warm oxygen are available.

While the meteorological conditions and the type of injury are beyond the control of the rescue organizations, they are able to increase the efficiency of communication, transport, evacuation and medical treatment. By doing this, the injured may be evacuated more quickly and safely, the quality of medical treatment may be improved, and the survival chances increased.

Technical aspects of mountain rescue

Rémy Henzelin

Organization for rapid and efficient rescue

Despite their grandeur and beauty, mountains can quickly become a nightmare for human beings. The effects of altitude, the isolation, the inhospitable surroundings, the difficulties of survival, and rapidly deteriorating weather conditions are all problems that mountaineers have to face. Added to these difficulties are avalanches of snow or serac, which remain the greatest danger in mountains.

In these surroundings, an accident which appears trivial can often take a dramatic turn if rescue is delayed. This is true of accidents occurring in any season, but is particularly valid in the case of winter avalanches. Avalanche accidents are often complex, and demand the greatest use of rescuers and resources within the shortest possible time. Moreover, if the first few minutes of the search are fruitless, further attempts may continue for several days or even weeks, entailing back-up problems, especially where large areas are affected. The most important thing to bear in mind in such a rescue operation is that for every hour that a person is buried alive in snow, his chances of survival decrease by 50% (**15**).

For this reason, the main aim of every rescue organization is to install the best possible ground and air rescue resources, in an attempt to minimize the time lapse between the accident and the patient's arrival at a suitable medical center. In avalanche accidents, a rescue operation consists of seven stages: the alarm; the implementation of rescue resources; the departure of the rescuers; the locating of buried avalanche victims, followed by their release; first aid and brief diagnosis; the evacuation of victims to a suitable medical center. While the success of an operation depends largely on the coordination and efficiency of the various stages, it is also subject to such variables as weather, terrain, quality of the snow, and severity of the accident. To achieve a high level of efficiency, each stage must be highly organized and well prepared, with the appropriate resources available, and fully integrated with the other stages so that no time is lost. Finally, every rescue operation must be placed under the control of a competent and experienced professional rescuer.

15

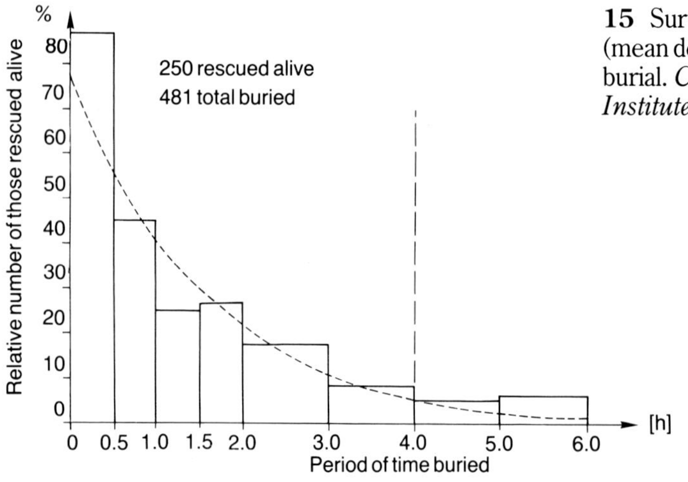

15 Survival percentage of buried avalanche victims (mean depth: 1.06 m) in relation to the duration of the burial. *Courtesy of the Swiss Snow and Avalanche Institute, Weissfluhjoch, Davos.*

The alarm

The alarm plays a decisive role in every rescue operation (**16**). Any delay greatly reduces the rescuers' chance of success and may even destroy it. This has always been particularly true of fire, sea and air disasters, and road traffic accidents. To this list can be added accidents in the mountains, which form a special category, characterized by the complexity and technical nature of the rescue resources employed, the conditions under which the rescuers work, and the increase in the number of incidents. Each year the statistics follow a rising curve, which now equals and, in some alpine regions, exceeds the number of traffic accidents.

Since the 1960s, alpine sports have been taken up by young people of both sexes, from mountainous and low-lying regions alike. Careful following of traditional ski runs has given way to off-piste downhill and cross-country skiing in any season, and climbing has never been so popular. Rashness has taken the place of caution, and what was unthinkable in the past is now becoming possible. Highly trained men and women, with the help of increasingly sophisticated equipment, can now be seen performing spectacular exploits. However, for each success there are many failures, as amateurs, seeing the ease with which a certain elite group performs in these surroundings, plunge in, underestimating or unaware of the dangers. Disaster soon follows.

The location of an accident and the prevailing weather conditions are different in every case,

16

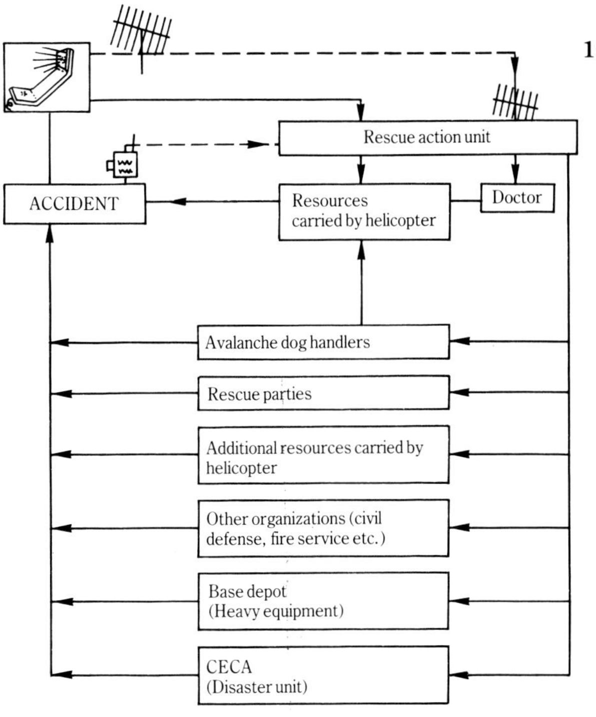

16 Relationship between the alarm and the rescue resources (Valais, Switzerland).

and need to be assessed very quickly by the rescue team. Even the most modern equipment can prove inadequate in what is essentially a race against the clock and the elements. It is here that the importance of the speed of the alarm is most evident.

Transmission of the alarm

Some years ago, any mountain hut not connected to the lowlands by the wire telephone system had no means of communicating with the outside world. When accidents occurred, the alarm was transmitted by land, sometimes by other mountaineers who witnessed the incident or saw distress signals from a distance.

The progress made in telecommunications over the years has played a decisive part in improving the speed of the alarm. In 1947, the first high-altitude refuges were fitted with radio telephones, powered by conventional batteries. Though not wholly reliable, these did provide some chance of calling for assistance and reducing the delay before the rescuers arrived. In 1988, new solar-powered installations of remarkable quality and reliability replaced them, and can now be found in most refuges (**17**).

However, the most important breakthrough has been the portable transceiver, which enables every climber with such a device to contact the rescue action unit directly, from anywhere within radio range (**18**). Though previously cumbersome and not very efficient, these appliances are now

17

18

17 A mountain hut keeper raises the alarm as he sees an accident.

18 Radio transmitter placed on the summit of Grand Combin in Switzerland, at a height of 4,314 m.

extremely light, of excellent quality and relatively cheap.

In Switzerland, certain rescue organizations now provide climbers and skiers with portable transceivers free of charge. This service was introduced following a tragic incident in 1974. Early in the morning of March 17, a guide and a dozen young cross-country skiers who had spent the night in a hut were debating whether to start out. Around 8 a.m., the weather cleared and they decided to continue on their route. At midday, the temperature increased and the snow became unstable. At 1.15 p.m., an avalanche occurred, and three of the skiers were buried in a mass of snow. Searches were begun immediately, but without success. The survivors were powerless and the guide decided to go for help. After two hours, he managed to reach the nearest refuge where he sounded the alarm. The rescuers reached the spot at 4.30 p.m. and recovered one of the victims, but all efforts to revive him failed. Night fell, and the group had to abandon the search and wait for the next day. The following morning, the two other bodies were recovered.

The incident occurred within eight minutes' flying time of a fully equipped helicopter base. Following this tragic event, guides were supplied with portable radios.

The contents of the alarm message

When raising the alarm, one must first decide if it is advisable to involve air rescue forces, taking into account the seriousness and urgency of the case, the accessibility of the scene of the accident by some other means, the weather conditions, and other such factors. A doctor's opinion may be required, and may be obtained either directly or via the rescue action unit.

The rescue action unit can be contacted either by telephone or by radio (**19** & **20**). The caller should indicate immediately his own name; the place from which he is calling; the telephone number or the radio frequency; and the incident which is the subject of the call.

The caller must then await instructions from the head of the action unit, who must be informed of the surname, first name and address of the person making the call; what has happened (fall, avalanche, crevasse, illness, disappearance, death, etc.); the date and time of the accident; the number of people involved and their names; the place, altitude and coordinates of the accident; and the condition of the people involved.

The head of the action unit should then be advised of the possibilities for a helicopter landing in the immediate vicinity of the accident, or whether a winch is necessary; a landing place may be prepared for the helicopter (*see* 'The landing site', p. 26). Details must also be given about the weather conditions, namely the force and direction of the wind, visibility, precipitation, and changing conditions.

The caller must specify whether there are any obstacles in the vicinity of the landing place or in the approach to it, such as electric cables or telephone lines. Any local assistance should be mentioned, and details given of any requirements, such as medical supplies, additional clothing and food.

It is particularly important that the telephone or radio used to contact the rescue unit is never left unattended, unless instructions are given to the contrary.

19

20

19 & 20 The rescue control center at Sion, Switzerland, can be contacted 24 hours a day, from any place in the Valais Alps, using a transceiver.

Rescue by helicopter

On November 23, 1946, a Fieseler *Storch* aircraft rescued the passengers of a US Army Air Force C-47 (*Dakota*), which had crashed four days earlier on the Gauli glacier in the Bernese Oberland, at a height of 3,200m (**21**). This was the first of many landings on glaciers, and marked the beginning of an era in which the techniques of flying in the mountains and of rescue flights with fixed wing aircraft were perfected.

Helicopters were not used in mountain rescue operations until the 1950s. At first, their use was restricted by the power of their engines, but with the introduction, in 1955, of Sud-Aviation's turbo-powered *Alouette II*, came great improvements. However, it was not until 1963 and the appearance of the *Alouette III* that the real revolution occurred. This machine is powerful, operational up to 5,000m, easy to handle, spacious and economical. It immediately became the ideal helicopter for mountain rescue. Its range of equipment has since been expanded and improved, and at present *Alouette III*s and *Lamas* perform thousands of rescues each year.

21

21 The US Army Air Force Dakota, which crashed in 1946 on the Gauli glacier in the Swiss Alps. *Courtesy of Swiss Air Rescue.*

The signs of distress

The international signs of distress in mountains are depicted in **22** & **23**.

22

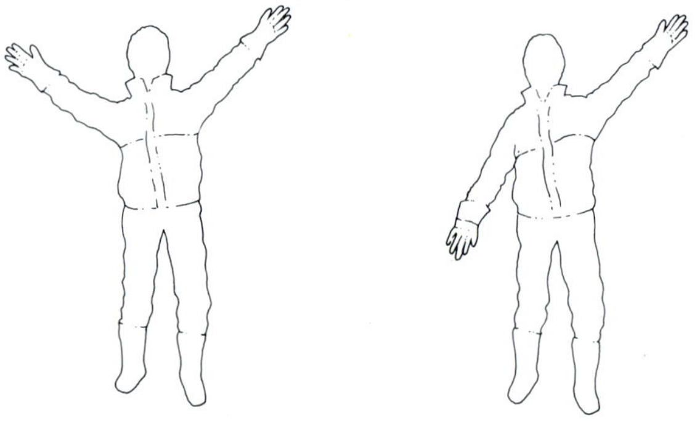

23

22 & 23 The international signs of distress are visible from a great distance. The arms held in a V-shape, so that the body forms a Y, signifies 'Yes! We need help; land here'. The arms held diagonally, away from the body, signifies 'No! Everything is all right; we do not need help'.

The landing site

First, a surface area 20m in diameter, and free of obstacles must be found. For the helicopter landing site, a flat or slightly raised area measuring 4 × 5m is required. Under no circumstances should the site be located on an area of depressed ground, as this entails the risk of collision between the main rotor or the tail rotor of the helicopter and the ground or bystanders. The landing site should never be marked out with skis or ski poles placed vertically.

If possible, powdery snow should be stamped down with skis and compressed over a surface area of $15 m^2$.

It should be remembered that helicopters always land and take off against the wind. Depending on the wind direction, the helicopter should, as far as possible, land away from any obstacles, such as wires, cables and pylons. Any such obstacles near the landing place should be marked with well-attached flags, handkerchiefs, etc.

No item that is likely to be blown away by the blast of the rotor, for example articles of clothing, sheet metal, boards or skis, should be present in the landing area. If the landing takes place near a building, the windows must be closed and the shutters attached.

Receiving the helicopter and directing the landing

Only one person must take charge of directing the approach and landing of the helicopter. Everyone else present should, wherever possible, stay about 30 m away from the landing point. If the terrain does not allow this, and the helicopter has to land near people, it is essential that they remain grouped together, crouched down, holding all equipment, such as clothing, bags, skis and ski poles, firmly and horizontally. No one should move until the helicopter is on the ground and permission has been given by the pilot or the flight technician. All those not directly involved in the rescue operation should stay away.

The person guiding the helicopter should stand with his back to the wind, on the edge of the landing area, preferably holding a handkerchief or some other piece of cloth to indicate the direction and force of the wind. Throughout the whole of the approach and landing, he must not take his eyes off the helicopter.

When the helicopter has approached and is landing, the person receiving it must stay where he is until it has actually landed. At the last minute, he should crouch down to avoid any danger of contact with the main rotor (**24**).

These actions are very important, as this person is the only fixed point of reference for the pilot in diffused light or powdery snow.

24

24 The helicopter must be approached only from the front.

Boarding with the rotor running

The pilot is always, except in improvised rescues, accompanied by a flight technician and, if necessary, a doctor.

The helicopter should be approached only when the main rotor is running at full speed or when it is stationary. It should never be approached when the rotor is at the braking or acceleration stage. Those boarding must wait for a sign from the pilot or flight technician before moving or approaching the machine. When approaching, dangerous areas of the helicopter should always be avoided.

If it is necessary to move around the helicopter, this should never be done from the rear as, in certain lighting conditions, the tail rotor can be completely invisible. Any movement around the helicopter should be from the front, remaining close to the cabin. When boarding, skis, ski poles, ice axes, aerials, radios, etc. must be held horizontally.

The courses of action

There are three courses of action that the helicopter pilot may take: a full landing, hovering, or winching.

Full landing

A full landing is the simplest course of action. It allows use to be made of the ground effect: the cushion of air that is established between the main rotor and the ground when the helicopter approaches its landing point, provided that the surrounding ground is also flat, and which ensures smoother landings and increases lift. A full landing also allows the rotor to be stopped if necessary, and permits people, particularly those injured, to board in comparative safety.

Hovering

No ground effect is present when the helicopter hovers (**25**), and consequently, 5% of the available power is absorbed; for example, at an altitude of 3,000m and a temperature of 0 °C, the weight which can be taken off within the ground effect is 2,000kg, whereas outside ground effect this is reduced to 1,900kg (i.e. minus 5% of the available power). In this situation, hovering therefore reduces by 100kg the live load that can be lifted, as compared with a complete landing on flat ground.

Winching

Winching (**26**) enables the rescue team to operate in steeply sloping areas, where no flat surface would allow the pilot to land or hover. However, like hovering, winching reduces by 5% the live load which can be carried. Furthermore, not every helicopter is fitted with a winch. Consequently, it is important to specify this requirement when raising the alarm. In addition, falling stones and ice may hit and damage the rotor when it is used near icy rock faces, especially during periods of thawing.

25

26

25 A difficult, but not impossible landing on one runner. *Courtesy of Swiss Air Rescue.*

26 When the helicopter is unable to land, rescue using the winch is the only possibility. *Courtesy of Swiss Air Rescue.*

Limitations in helicopter use

The helicopter has certain limitations, which restrict its use in mountain rescue operations. With altitude, its overall performance worsens, and its lifting ability, and consequently its live load, is reduced. To avoid danger, these limitations must be respected by the pilot, who should calculate an accurate weight estimate before each take-off. The pilot should take only the amount of fuel necessary for the rescue operation, together with the compulsory reserve of 20 min, so that he is left with the greatest possible margin for live load, crew, equipment required for the rescue operation, and the victim or victims to be evacuated. In some situations, several round trips may be necessary to completely evacuate the accident area.

Restrictions in helicopter weight are particularly important in rescue operations requiring medical assistance, in which bulky equipment weighing up to 30 kg is carried. For this reason, systematic medical assistance for every helicopter rescue is not indicated and a more flexible approach is preferred, in which the needs of each individual case are assessed.

In the mountains, the pilot has to fly 'by sight', i.e. using external visual references to find his bearings. Despite the great technological progress of recent years, helicopters involved in mountain flights are still without instrumentation that will enable them to navigate freely in mist, white light, precipitation, and other weather conditions affecting visibility (**27**).

Night flying is particularly problematic. On very dark nights or when the sky is overcast, the helicopter pilot is entirely without visual references; these exist only when the sky is clear during the period between the waxing and the waning half-moon. Night flying can pose serious problems for single turbine helicopters. If the propulsion unit breaks down, the pilot has practically no chance of carrying out an emergency landing, in autorotation, on uneven ground. Moreover, other seemingly trivial incidents, such as the failure of a searchlight, may have serious consequences.

At night, it is difficult to obtain reliable information regarding prevailing weather conditions at the scene of the accident and also on the flight route which the helicopter will be following. In addition, when landing in darkness, it is practically impossible to estimate the strength and direction of the wind, and it is equally difficult to assess the quality of the landing site. Furthermore, all obstacles are invisible. With no visual clues to guide him, the pilot may become completely disorientated, and possibly lose control of the helicopter.

For these reasons, night flying demands perfect knowledge of the sector by the pilot. Even then, the chances of success for a nocturnal search flight are very small, and any such operation will place the rescuers' lives at great risk. Generally, the only nocturnal helicopter rescue that can be considered is the evacuation of people whose lives are at serious risk, and who have been able to prepare an illuminated landing area.

27 Bad weather conditions are common during rescue operations. *Courtesy of Swiss Air Rescue.*

Determining factors in helicopter rescue operations

The quality of the crew

- The pilot.
- The flight assistant.
- The doctor/guide.

} Homogeneity = efficiency.

The equipment

- Technical equipment.
- Medical equipment.

The weather conditions

- Mist = poor visibility.
- Precipitation.
- Wind and terrain = turbulence.
- White-out = absence of visual references.
- Icing up.
- Temperature.

The terrain

- The pilot's and the crew's knowledge of the terrain.
- The nature of the ground.
- Uneven terrain and the effect of the wind.
- Glaciers and crevasses.
- Obstacles, such as cables and electricity lines.
- Accessibility: the presence of rock faces, slopes, etc.
- The danger of falling rocks, and of avalanches.
- Altitude: the helicopter's performance and the effects of the lack of oxygen on the crew.
- The possibility of turning back if weather conditions deteriorate.

Communications

- The quality of communications within the rescue unit.
- The efficiency of the rescue unit.

The extent of the risks taken by the pilot

Every rescue operation involves risks. These may appear minor in good weather conditions, but can rapidly become serious if the weather worsens, or if a rescue operation must be performed at night. The pilot then faces a difficult decision: should he take increasingly large risks and, in so doing, impose them on his crew, or should he abandon the operation? Following some serious accidents and near disasters, a certain code of conduct has emerged, based on a sharing of responsibilities. The mountaineer must be aware of the risks that some of his actions entail, and be able to take responsibility for his own safety as far as possible in difficult circumstances, i.e. when the helicopter cannot intervene without placing the safety of the rescuers at risk. Unfortunately, this sharing of responsibilities is not always understood, either by the mountaineers or by their families and friends.

Meteorology in the mountains

While weather forecasting plays a very important part in aviation in general, it is of limited use to helicopter pilots in the mountains, especially in emergencies. In the mountains, the pilot is largely on his own, and to conduct a rescue operation safely, he can count only on the reliability of his machine, and on his knowledge of the terrain and of certain elements that are peculiar to mountainous regions. Meteorology is one of the most important aspects of this knowledge, and forms the basis of the technique of mountain flying. The most elementary of its laws are described below.

The following phenomena rarely appear in isolation, and are nearly always combined with one or more others.

Thermal effects of the terrain: valley and mountain winds

28

28 The valley wind.

29

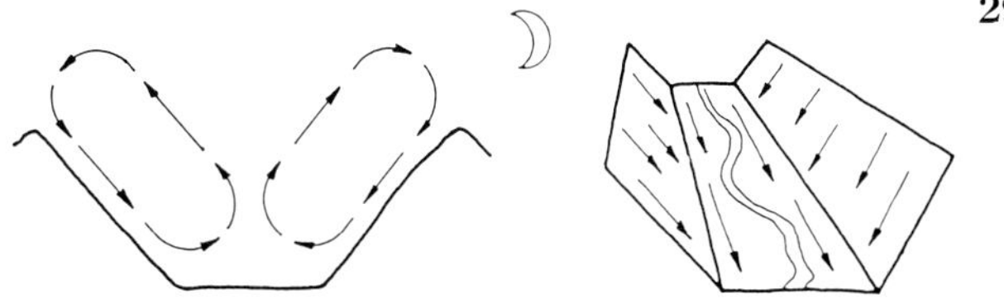

29 The mountain wind.

During a twenty-four hour cycle, a valley undergoes variations in pressure, caused by the effects of sunshine during the day and by radiation from the ground at night.

In the morning, the layer of air lying nearest the sunlit slopes is warmed through contact with the ground. This air becomes lighter than the surrounding air, rises and produces a circular movement up and along the slopes of the valley (convection). In mountainous regions, this begins about 30 min after sunrise, and is called the valley wind (**28**). During the day, depending on the humidity of the air, groups of cumuli which extend vertically may begin to form.

At about 17.00, the sun begins to set and the ground begins to cool through radiation. As it comes in contact with the ground, the air drops in temperature and consequently becomes heavier. An inverse circular movement is then established, about 30 min after sunset; this is the mountain wind (**29**).

The active layers of these winds are about 300 m thick and, during the day, can reach a force of 25–30 km/h in larger Alpine valleys, such as the Rhône, the Rhine and the Ticino valleys. The mountains and valley winds can be cancelled out or strengthened by strong air currents.

Mechanical effects of the terrain: turbulence

30

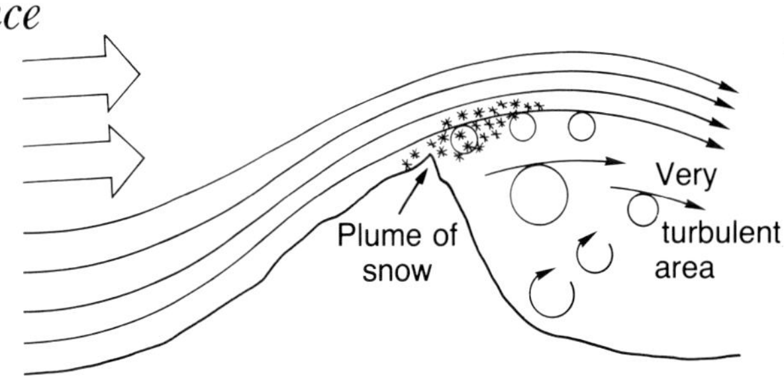

31

30 & 31 Strong wind turbulence and its effects.

The lower layers of the wind, which lie next to the ground, are inevitably influenced by the terrain and follow the laws of aerodynamics.

The laminar flow of a current of air over a smooth surface, such as that of a lake or a plain, becomes turbulent if obstructed by obstacles on the ground. If the ground's surface is relatively smooth, a current of average strength, blowing up the windward slope of a mountain, will not be obstructed. However, on blowing down the leeward slope, the wind will rarely follow the terrain, becoming turbulent and forming folds of varying dimension.

In very strong winds, air currents obstructed by ridges 'take off' from the mountain side and continue blowing up past the summit of the windward slope, carrying snow which forms a white, powdery plume, before descending. These plumes of snow are generally clearly visible and indicate violent winds and very strong turbulence (**30** & **31**). In these conditions, very rapid circular air movements called 'rotors' may develop, the axes of which vary with the terrain, and which may move in any direction.

Flying in areas such as these, where the turbulence is particularly violent, is extremely dangerous and must be foreseen by the pilot and avoided at all costs. In some cases, the intensity of the turbulence may cause permanent damage to the helicopter.

The valley effect

Narrow valleys also produce turbulence, through the rising air currents at their heads and the descending air currents at their exits. When these air currents circulate perpendicularly to the valley, strong turbulent winds may result as the currents blow up the windward slope, and descend along the leeward slope.

Anabatic (rising) winds

As we have seen, air currents in the mountains are governed by the laws of aerodynamics. Therefore, by observing the nature of the terrain and being aware of the basic principles governing the flow of air currents, it is theoretically possible for those awaiting rescue to determine the most advantageous landing site for the pilot, i.e. the spot that will best allow him to benefit from anabatic winds.

On the windward slope of a mountain, the speed of the air may be broken down, according to the rule of the parallelogram of speeds, into a horizontal component, giving the translation speed, and a vertical component, called the vertical speed. These anabatic winds form a band above the windward slope, which may be used by the pilot to improve the performance of his machine.

The area of influence of an obstacle

The layer of air which is disturbed by an obstacle constitutes that obstacle's 'area of influence'. This area varies greatly in height and depth, and depends on the nature of the terrain (whether a single mountain or a chain), on the obstacle's height, the atmospheric stability, the direction and force of the wind, etc. However, one can estimate the area of influence to be approximately one-third of the obstacle's height.

The undulatory effect

The terrain's influence is not restricted to those layers of air closest to the ground. Under certain conditions, such as the presence of a foehn or a north wind, the upper layers are also disturbed. At altitude, the undulatory currents created by the profile of the terrain produce formations of lenticular altocumuli, while in the lower layers of the atmosphere, rotors can be observed, the presence of which is revealed by stationary cumuli (**32**). Inside a fast-moving rotor, air movement may be very steady. However, the rotor's bordering regions are extremely turbulent. These air currents, which are often strong and gusty, can become dangerous. They may also be encountered in the form of violent turbulence in clear sky, at a distance of several hundred kilometers from their origin.

32

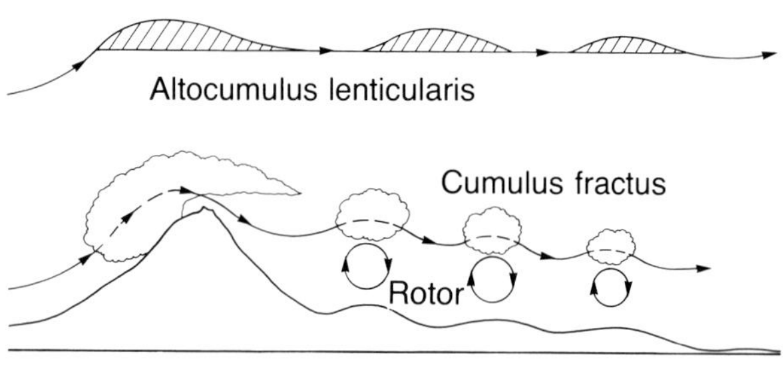

32 The undulatory effect associated with the presence of rotors.

The Venturi effect

Uneven terrain, windswept passes and narrow valleys can act as Venturi tubes and produce similar effects. In such areas, the following may be noted: stronger winds, whose speed may double; increased turbulence; and a drop in atmospheric pressure, due to the increased speed of the air current (Bernoulli's law). This drop in local pressure influences altimeters, which then show an inaccurately high altitude.

Physical effects of terrain: the foehn

When a current of moist air is forced to pass perpendicularly through a high mountain range, it cools as it rises up along the slopes. On reaching a certain height, the relative humidity increases, condensation occurs and produces considerable precipitation in the form of rain or snow (moist adiabatic gradient). When the mass of air reaches the summit of the mountain, at which point it has lost most of its moisture, it blows down the opposite slope, gradually becoming warmer and producing a warm, dry, and very turbulent air current in leeward valleys (dry adiabatic gradient) (**33** & **34**).

On leeward slopes, the foehn is generally so strong that it causes a downward pressure and rotors to form, against which a helicopter finds it extremely difficult to fly. A downward vertical speed of 10 m/s is found frequently. In some particularly exposed valleys, this may reach 15 m/s or even more. A helicopter with a rising speed of 4–5 m/s will therefore drop several hundred meters per minute, if caught in such downward pressure, and, if it fails to move to a safer area immediately, will be in danger of crashing to the ground.

When flying through a series of violent eddies, such as those produced by a foehn or a strong north wind, both the crew and the passengers are jolted about. In addition, strong turbulence may cause the machine's admissible load factor to be exceeded, and distort or even break the helicopter's airframe and moving parts. Such situations call for great caution, and may even entail abandoning the rescue operation.

33

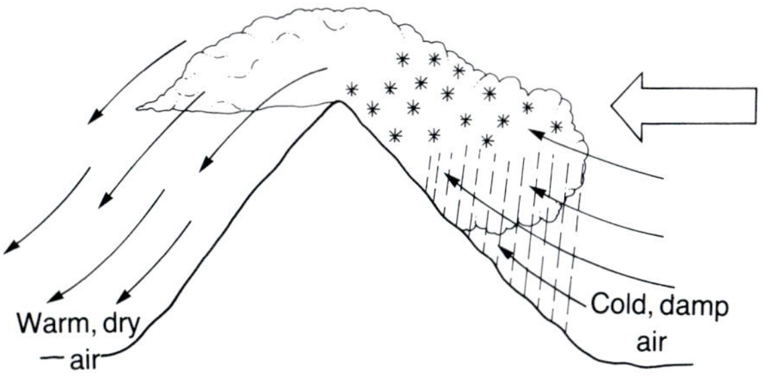

34

33 Factors involved in the creation of the foehn.

34 Environment in the Alps prone to foehn.

Medical assistance in mountain rescue

Frédéric Dubas

The development of medical assistance in mountain rescue

35 The Champéry rescue team in Switzerland, in the early twentieth century.

The development of mountain rescue has paralleled that of mountaineering. The first known structured mountain rescue organization was set up by the monks of the Great St Bernard, whose monastery, founded in the tenth century by St Bernard of Menton, is situated in the pass of the same name. There, assisted by their dogs, the monks brought help to many pilgrims lost in the mountains on their way to Rome. This same route between the northwest and the south of Europe, in use for over 2,000 years, became famous when it was crossed by Napoleon Bonaparte's army in 1800.

In view of the increasing popularity of mountaineering, inhabitants of the mountains eventually had to set up rescue organizations. These initially only involved the villagers, but have since been progressively reinforced by rescue units set up by the national mountaineering clubs, which were founded during the second half of the nineteenth century (**35**). These volunteer organizations are currently still in force in most countries, but only operate when weather conditions will not allow an airborne rescue.

Since the end of the Second World War, aircraft, and later, helicopters (**36**), have rapidly developed into invaluable tools in mountain rescue. Whereas a mountain rescue operation once consisted largely of bringing corpses down to the valleys, the speed of airborne rescue now enables the injured and the sick to receive immediate medical attention.

36 This Bell 47–J was the first helicopter used for mountain rescue in Switzerland. It was offered to Hermann Geiger in 1957. *Courtesy of Swiss Air Rescue.*

In most industrialized countries, rescue with medical assistance has become the generally accepted practice. In large towns, ambulances nearly always carry medical resources, and in helicopter rescues, for example at motorway accidents, a doctor, together with sophisticated medical equipment, is often on board. The presence of a doctor at the scene of the accident was quickly sought by mountain rescue groups. A Swiss rescue specialist, Hermann Geiger, flying a Piper Cub fitted with skis, was the first to call on doctors when evacuating workers from worksites in the high mountains (**37**), or rescuing climbers.

Initially, this medical assistance was occasional, and the doctors had no special training in mountain medicine. The foundation of the Swiss Air Rescue Guard (REGA) by Fritz Bühler in 1952 resulted in the organization of a general rescue system in Switzerland. From 1973 to the present date, doctors have regularly been employed by the Air Zermatt company in helicopter operations.

In 1976, a major telecommunications network was set up throughout Switzerland to facilitate radio communications throughout the country and, in particular, in the rugged massif of the Alps. The prognosis of mountain accidents has now been transformed by the immediate medical care made available to the sick and injured, by greatly improved telecommunications networks and airborne rescue techniques.

37 

37 Hermann Geiger (1914–1966), here on the left, was the pioneer of mountain air rescue in Switzerland. *Courtesy of Swiss Air Rescue.*

The situation outside Switzerland

- **In France,** the Platoon of High-Mountain Gendarmes (PGHM), based in Chamonix, has regularly taken a doctor on its helicopter rescue operations since 1975, conducted in collaboration with the High-Mountain Military Academy (EHM) of Chamonix. The Urgent Medical Assistance Service (SAMU 38) of Grenoble has also been active in developing this practice. The SAMU, a French national organization, has played a major role in the development of medical assistance in mountain rescue throughout the Alps and the Pyrenees.
- **In Italy's** Aosta Valley, a mountain rescue organization was set up in 1988. Since this date, all mountain rescues have included medical assistance, in the form of general practitioners or volunteer anesthetists. The first doctors' training course in mountain rescue took place in 1989.
- **In other European countries,** rescue operations with medical backup are gradually becoming more common.

The purpose of medical assistance in mountain rescue

The Groupe d'Intervention Médicale en Montagne (GRIMM: Medical Action Group in the Mountains) was set up in 1981. Its aims and the reasons for its foundation are outlined below.

(i) The provision of medical assistance in mountain rescue involves bringing a doctor, who can immediately deal with the patient's injury or sickness, to the scene of the accident. The patient may then receive optimal care, and the choice of an appropriate medical center may be made immediately, thereby avoiding pointless detours. The patient's condition must be assessed and, if possible, stabilized, to allow safe transportation (**38**); the doctor may use resuscitation, intubation, the insertion of an intravenous drip, an analgesic, or even narcotics. During the flight, little medical assistance can be given, due to the cramped conditions inside the helicopter.

(ii) The presence of a qualified doctor at the scene of the accident strengthens the rescue team, allowing its members to focus all their attention on the technical, rather than the medical, aspects of the rescue.

(iii) Including a doctor in the rescue team can also give non-medical rescuers the chance to gain paramedical training.

(iv) The doctor is one of the major links in the rescue chain. He is best qualified to give details of the patient's injuries or sickness to the medical center and, inversely, is also in a position to inform the rescue team of 'their' patient's subsequent progress.

(v) Knowing a doctor was present at the scene of the rescue reassures the family of patients who have died that everything possible was done to save them. This is an important psychological aspect.

The doctor must be perfectly integrated in the rescue team; a team is only as strong as its weakest member. Therefore, the doctor must be able to assist in all aspects of the rescue.

38

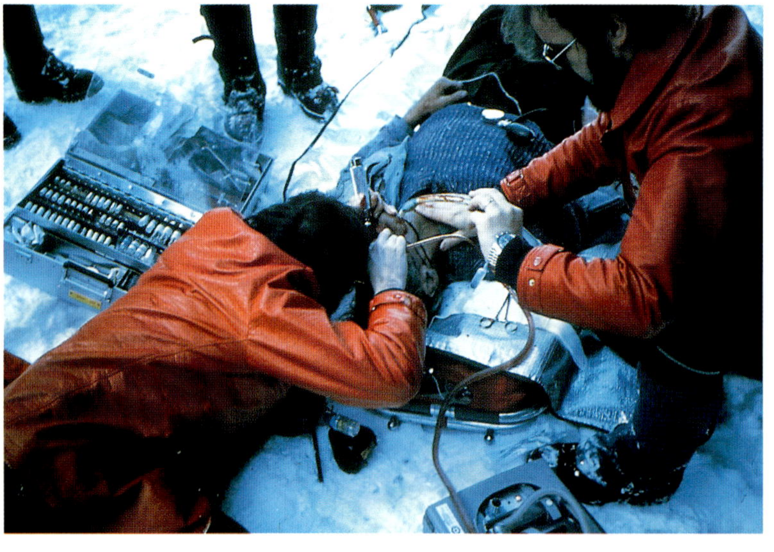

38 Resuscitation and, in particular, intubation can be difficult to perform in the field. *Courtesy of Swiss Air Rescue.*

The GRIMM action pack

Each member of a GRIMM team involved in a rescue operation carries the following equipment in his pack:

Pocket 1
Complete intubation kit (laryngoscope, several Portex tubes, Crile's clamps, Magill's forceps, Guedel tubes, spare equipment, syringes, Heimlich valves, etc.).
Pocket 2
Bag 1: aspirating equipment. Bag 2: ventilation equipment (oxygen bag and mask). Bag 3: perfusion equipment (glucose solution, physiological solution, plasma-expander, intravenous tubes).
Pocket 3
Flashlight with spare battery and bulb.
Pocket 4
Bandages, surgical knife, syringes, tourniquet, eye bandages.
Pocket 5
Intravenous fluids.
Pocket 6
One inflatable leg splint.
Pocket 7
One snow shovel handle.
Pocket 8
List of the equipment.
Pocket 9
Empty reserve pocket.
Pockets 10 & 14
Drugs: prednisone, lignocaine, atropine, sodium bicarbonate 8.4%, calcium, terbutaline, pentazocine, verapamil, propanolol, frusemide, digoxin, diazepam, metamizole, adrenaline (epinephrine), methylergometrine, nitroglycerin, nose drops.
Pocket 11
One snow shovel.
Pocket 12
One stethoscope.
Pocket 13
Bandages: elasticated dressings, etc. One thermometer for epitympanic measuring of core temperature (*see* p. 209), plastic bags for vomit and hyperventilation, for example.

The pack should also contain one miniscope (cardiac monitor) (*see* p. 209), one horizontal net for winch rescue, one extension cable for the winch, and one locking carabiner.

Organization

The doctor's involvement

The doctor's involvement depends on the precision of the information obtained from those who raised the alarm; the seriousness of the victim's condition; the number of air rescue resources available.

The decision to involve the doctor is a difficult one. In some countries, such as France (thanks to the Urgent Medical Assistance Service (SAMU)), the alarm is transmitted directly to the doctor, either by telephone or by radio. The doctor can then ask the necessary questions and choose the most appropriate course of action, depending on the resources at his disposal, such as rescue units, ambulances, ambulances with medical equipment, helicopters, etc.

In other rescue organizations, the doctor is invariably on board the helicopter, whether involved in the rescue or not. In this way, the doctor's presence is always guaranteed, should it be necessary, though he is frequently used as an extra flight technician.

When several helicopters are available, an order of priority may be assigned to each machine. For example, in an avalanche rescue, the first helicopter will take rescuers and avalanche dogs, with the doctor following in a second helicopter and becoming involved only at a later stage, when the victims are being freed. However, where there is a victim suffering from heart trouble, in a problem-free area, the doctor must be on board the first helicopter to reach the site.

The availability of the doctor

A rescue doctor in the high mountains must be available 24 hours a day, 365 days a year. There are two types of rescue operation in which he may be needed: air and ground.

Air rescue operations

Nowadays, air rescue operations are the most common. In the Alps, 95% of mountain rescues are by helicopter. In this type of operation, the doctor must be immediately available at all times. In fact, the helicopter must be able to take off within 10 minutes of receiving the alarm (**39**).

Depending on the country and the region, the doctor may be engaged professionally by the helicopter company, or may run an independent mountain rescue service and be employed as required. Both these situations have advantages and disadvantages.

Professional rescue doctors are readily available and are very familiar with the crew and the equipment on board the helicopter. However, in the event of a disaster, the number of professional rescue doctors available is limited. In addition, they tend to become dissatisfied because of the lack of intellectual stimulus, which leads to a lack of motivation and, consequently, a reluctance to stay in the position for extended periods. These doctors have little or no contact with ground rescue teams, and are not involved in any ground rescue operations.

Independent rescue doctors have the following advantages: the number of independent rescue doctors available is virtually unlimited; they have generally worked as team members for some time, becoming progressively more efficient and more familiar with both their crew and with ground rescue teams; they may be called on to assist in ground rescue operations. However, these doctors are less readily accessible and, initially, their competence is difficult to assess.

39

39 The helicopter rescue crew must be a homogenous team (pilot, co-pilot, mountain guide and physician), working together for maximum efficiency. *Courtesy of Air-Glaciers.*

The author prefers the independent doctor system, and feels that the motivation, long-term collaboration and integration in both air and ground rescue teams are advantages of prime importance. Under this system, each doctor deals with the incidents which occur in his region, and the helicopter picks him up on its way to the site of the accident. This allows the doctor time to leave what he is doing, to prepare his equipment and to go to the place where the helicopter will pick him up.

Ground rescue operations

Rescue doctors based in the high mountains must be experienced in ground rescue work. In the Swiss Valais, it is the regional doctor, a fully integrated member of the rescue unit, who takes part in ground rescue operations (**40**).

40

40 For about 90 days a year, bad weather conditions prevent flying in the Alps. Rescue operations are then performed by ground rescue teams.

Training

A high-mountain rescue doctor must acquire both technical and medical training.

Through his **technical training**, he must become an experienced climber, able to move and to survive in the mountains, without being a burden or a source of concern for the other rescuers; he must be able to participate in rescue operations as fully as any other member of the team, becoming skilled in the use of rescue equipment, able to work while suspended by a winch above rock faces, and he must become familiar with all aspects of helicopter rescue.

Through his **medical training**, the doctor must acquire a good basic knowledge of first aid, and of surgical and medical resuscitation. He must therefore be skilled in intubation and resuscitation in difficult conditions. His medical knowledge must be demonstrated and tested during annual training courses. It is regrettable that the teaching of mountain medicine is not included, even on an optional basis, in the syllabuses of the Medical Schools in those countries with high mountains.

Equipment

Rescue doctors based in the high mountains must have the following equipment:

- **Technical:** full mountain equipment, selected according to the season, including harnesses, skis, crampons and ice axes.
- **Medical:** various action packs of equal value have been produced by a number of groups in different countries. They contain the necessary equipment for optimal medical care in the high mountains (**41**) (*see* GRIMM action pack, p. 37).

In addition, various articles must be available for both ground and helicopter rescue operations, namely insulating mattresses, stretchers with runners, inflatable splints, defibrillators, oxygen bottles, bronchial aspirators, etc. (**42**).

41 

41 Rescue packs must be fully equipped, lightweight, and comfortable to carry on the back. Pictured here is the GRIMM action pack.

 42

42 Special equipment has been developed for mountain rescue. Pictured here is the horizontal net, used where the victim has trauma to the vertebral column.

Medical training for mountaineering professionals

Caretakers of mountain refuges, mountain guides, ski instructors and ski slope patrollers are often the first to arrive at the scene of an accident. They are therefore regarded as vital colleagues of the rescue team, fulfilling the role of paramedics in the rescue operation. For this reason it is essential for them to be provided with adequate training in mountain medicine.

For many years, GRIMM has, via its members, been organizing courses in mountain medicine for the various professional groups mentioned above. The syllabus is as follows:

(i) Principles of resuscitation and first aid.

(ii) Knowledge of mountain medicine (altitude, hypoxia, cold, sun, exhaustion, nutrition and treatment of the various minor injuries which may occur in mountaineering).

(iii) Establishment of good relations between the doctors and these valuable assistants. This can often allow problems to be settled by telephone.

(iv) Discussion of cases experienced by those present, resulting in a mutual increase in knowledge.

43

PART I

CLIMATIC CONDITIONS AND ASSOCIATED PATHOLOGIES

43 Ama Dablam in the Himalayas of Nepal (altitude 6,883 m). *Courtesy of N. Hubert.*

Altitude

Definition and parameters

Paolo Cerretelli

For decades, high altitude has been one of the most challenging environments, not only for the mountaineer, but also for the physiologist. The interest originates mainly from the lower partial pressure of inhaled oxygen ($P\text{IO}_2$), followed by a drop in arterial O_2 pressure ($P\text{aO}_2$), and from the effect of the latter on the oxygen transport cascade from the ambient down to the mitochondria (**44**). To maintain the critical O_2 pressure at which cells may function normally, a series of adaptations (acclimatization) need to occur in the body. These consist of hyperventilation, changes in the acid–base balance, increase in the blood O_2 carrying capacity, changes to the shape of the oxyhemoglobin dissociation curve, as well as changes in cardiac output and to the peripheral circulation.

In evaluating the adaptive reactions of an acclimatized lowlander to high-altitude exposure, the reference condition is that of 'natural acclimatization', typical of high-altitude natives. Steps in the adaptive process range from short-term reactions to acute hypoxia and to so-called 'acquired acclimatization', a condition in many respects similar to that of the altitude native.

44

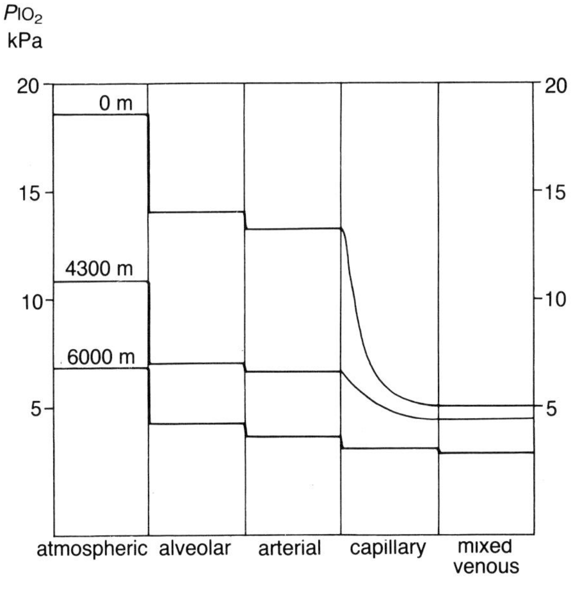

44 The oxygen transport cascade from the ambient down to the mitochondria (described by C. Houston, 1980).

The main physical variables concerning the altitude environment

Barometric pressure

Barometric pressure (P_B, mmHg, Torr or kPa*) is the primary effect of the earth's gravitation on the atmosphere and, therefore, at a constant temperature, it decreases exponentially with the distance from the earth's center. **45** shows a plot of P_B as a function of altitude, based on the Standard Atmosphere values of the International Civil Aviation Organization (ICAO), which are an attempt to represent average atmospheric conditions in temperate latitudes. Also shown in **45**, together with the ICAO curve, is the curve obtained from the formula proposed by Zuntz (Pugh, 1957).

Direct measurements, taken particularly in the Himalayan range during the summer, have been shown to correspond more closely to the pressures calculated from the Zuntz formula, rather than to the ICAO curve (Pugh, 1957; Dill and Evans, 1970; West *et al.*, 1983). Latitude plays an important role in these recordings. At latitude = zero, P_B is greater than predicted, owing to the very large mass of cold air accumulated in the stratosphere above the equator (Brunt, 1952). The season also affects the recorded barometric pressure. The seasonal (January vs. July) pressure differential for the northern hemisphere at the latitude of Mt. Everest is approximately 1.33 kPa (West *et al.*, 1983). As a practical consequence, both in the pre-monsoon (May) and post-monsoon (October) period, the recorded P_B on the Everest summit is some 2.0–2.7 kPa higher than that predicted from the shape of the stratosphere. The value recorded by the 1981 American Scientific Expedition is 33.7 kPa. The physiological significance of such a pressure differential in climbing Mt. Everest has been discussed by West *et al.* (1983). The practical conclusion is that it would be impossible for a climber to reach the summit without supplementary oxygen, if the barometric pressure there were as low as predicted by the ICAO curve.

*1 mmHg ≈ 1 Torr ≈ 0.133 kPa.

45

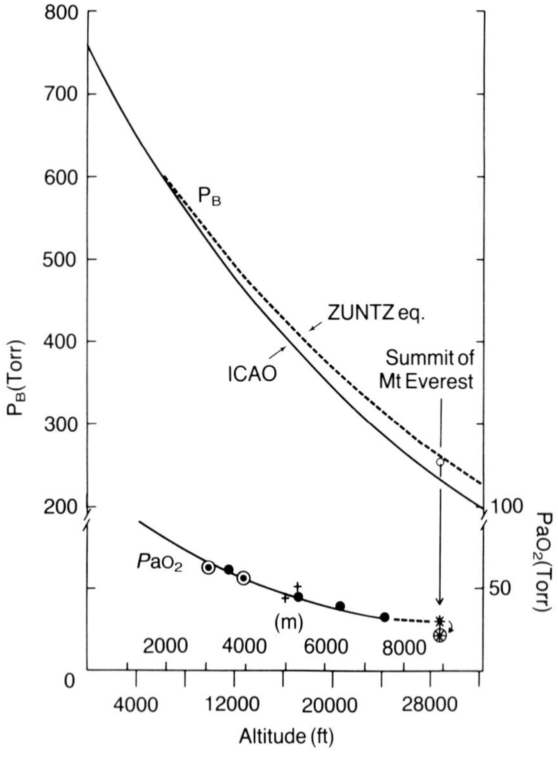

45 Barometric pressure (P_B) and alveolar O_2 partial pressure (PaO_2) as a function of altitude. The asterisks designate the expected PaO_2 on the summit of Mt. Everest for two different P_B values, i.e. that based on the International Civil Aviation Organization (ICAO) chart, and that calculated from the equation of Zuntz. The actual value represented by the open circle is 33.7 kPa. *Modified from Cerretelli, 1980.*

Air density

The density (ρ) of a gas (m/l^3) decreases with increasing altitude. Rather than its absolute value, it is the relative density of air, with reference to that at sea level (ρ/ρ_O), that is of most interest to respiratory physiologists, because of its effects on breathing mechanics. **46** shows a plot of ρ/ρ_O versus altitude at the indicated prevailing temperatures.

The changes in airway resistance as a result of varying the elevation are not those that would be expected from the changes in air density, for example, as predicted by the equation of Varène *et al.* (1967). On the basis of the reduced ρ, airway resistance should, in fact, be 17% less at 61 kPa than at sea level. In lowlanders exposed to altitude, the actual average decrease was found to be considerably less than that predicted (0.28 vs. 0.30 kPa at $V \sim 1$ liter/s) (Cruz, 1973). Furthermore, paradoxical reactions were observed. It is probable that hypoxia and/or hypocapnia, by causing airway narrowing and thus increasing airway resistance, counterbalances to a variable extent in different individuals the effects of decreased air density. This could explain the wide scatter of the data and the occasional finding of increased airway resistance, despite decreased air density. As a result of decreased air density, maximal voluntary ventilation (MVV) may be markedly increased in lowlanders upon prolonged exposure to altitude.

Because of increased pulmonary ventilation ($\dot{V}E$), work of breathing (W_b) at rest is higher at altitude than at sea level. When corrected for the increase in $\dot{V}E$, W_b of lowlanders is still significantly greater at altitude than at sea level, mainly because of an increased flow-resistive component. This result is the opposite to what would be expected from the reduced air density, and is probably the result of hypoxia and/or increased tissue frictional resistance. The ventilatory response ($\dot{V}E$ BTPS) of acclimatized lowlanders to increased metabolism ($\dot{V}O_2$) is greater as P_B decreases to at least 40 kPa (7,400 m), but $\dot{V}E$ STPD tends to be remarkably constant at any given altitude.

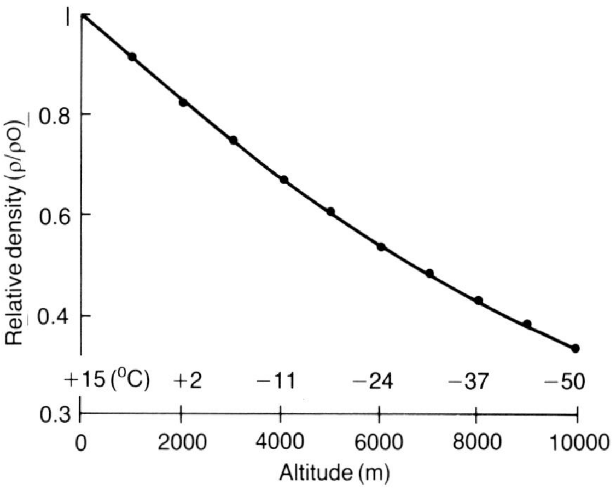

46 Relative air density (ρ/ρO) as a function of altitude. Prevailing temperatures (°C) at various altitudes are also shown.

Air absolute humidity

Air humidity is the mass of water vapor present per unit volume of atmosphere. It decreases with altitude as a result of the temperature drop. Dryness of inspired air coupled with altitude hyperventilation increases water loss through the airways, particularly during muscular exercise. It may be calculated that water loss through respiration during a 10 hour walk at average speed with a $\dot{V}E$ BTPS of 50 liters/min, when performed in an environment at 0°C ($PH_2O <$ 0.53 kPa), may reach 1.5 liters. This accounts for the body dehydration observed during altitude climbing.

Performance and metabolism at high altitude

Paolo Cerretelli

The maximal aerobic power ($\dot{V}O_2max$) of man drops as barometric pressure (P_B) decreases at constant FIO_2 and/or with decreasing O_2 partial pressure in the inspired air, PIO_2 (**47**). For a given decrease in PIO_2, the reduction in $\dot{V}O_2max$ is approximately the same in acute hypoxic subjects, in acclimatized lowlanders, and in altitude natives. The maximal aerobic power on the summit of Mt. Everest has been estimated at about 25% of the sea level reference value, i.e. 1 liter/min or threefold the resting level. A sudden rise in PIO_2 to 50.7 kPa was found to raise the $\dot{V}O_2max$ of lowlanders acclimatized to 5,350 m, from 70% to only 92% of the control sea level $\dot{V}O_2max$ value, despite a 37% increase of blood Hb concentration.

In **48**, as maximal cardiac output ($\dot{Q}max$) at 5,350 m was only 10–20% lower than at sea level, the observed persisting limitation of $\dot{V}O_2max$ upon restoration of normoxia can be attributed to a peripheral reduction in O_2 transport, due to impaired muscle perfusion and/or utilization by the tissues.

47

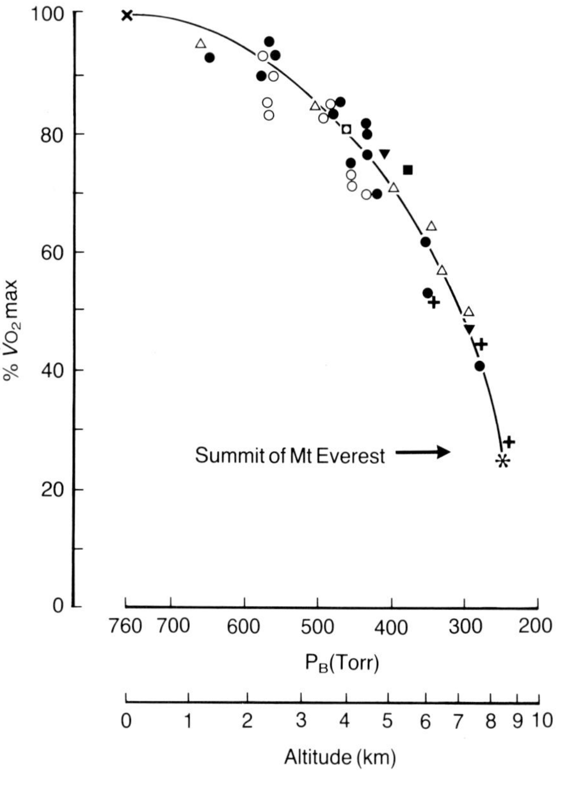

47 $\dot{V}O_2$ as a function of altitude.

Open symbols = acute hypoxia
Filled symbols = chronic hypoxia
\+ = operation Everest II
* = summit of Mt. Everest

48

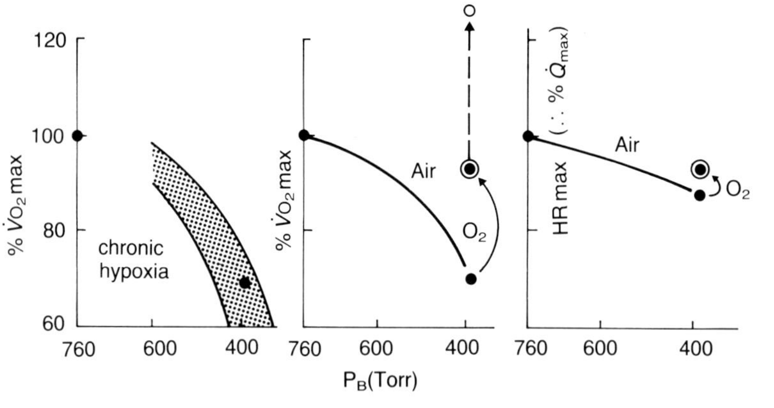

48 Effect of raising PIO_2 on the $\dot{V}O_2$ max of lowlanders acclimatized to 5,350 m and above.

Left: range of $\dot{V}O_2$ max decrease as a function of altitude; the full dot is the average of 10 subjects.

Center: the effect of raising PIO_2 on the $\dot{V}O_2$ max of the investigated subjects. The open circle at the top is the value calculated on the basis of extrapolated $\dot{Q}max$ and measured Hb concentration.
Right: HRmax changes as a function of altitude and a sudden increase in PIO_2.

Recent experimental findings, which support the above hypothesis, have been obtained on subjects acclimatized for five to eight weeks to altitudes of 5,200 m and above, upon return from the 1981 Swiss Expedition to Mt. Lhotse Shar (8,398 m): firstly, a 30–40% reduction in muscle blood flow was noted during two standardized submaximal cycloergometric efforts; secondly, there was a 10–15% decrease in muscle mass, as assessed by computer tomography of the thigh; and thirdly, a reduction in the activity of some enzymes of the TCA (Krebs) cycle was noted. These observations explain why the drop in $\dot{V}O_2$max may be similar in acute and chronic hypoxia, up to 7000 m. In the former condition, $\dot{V}O_2$max is limited by O_2 transport (reduced Hb oxygen saturation), and in the latter condition, by a muscular defect, as the O_2 transport is partly improved by increased hemoglobin concentration, whereas maximal cardiac output is still relatively well preserved.

In lowlanders acclimatized to ~ 5,500 m, resting blood lactate was found to be unchanged, but the maximal lactate concentration observed after exhausting exercise was reduced to about half the sea level control values (**49**); a similar decrease in the 'apparent' buffer value of the body was also noted. The latter parallels the drop in bicarbonates in plasma and in extracellular space. In acclimatized lowlanders, the rate of lactate removal from blood after a supramaximal exercise was found to be the same as at sea level. However, lactate diffusion from the muscles into the blood appears to be initially slower, probably because of the decreased muscle perfusion. The maximal muscular power ($\dot{W}$max or peak power) assessed during a standing high jump on a force platform using both feet, is not affected by chronic hypoxia for up to three weeks' exposure to 5,200 m. However, longer exposures lead to a ~ 25% drop in $\dot{W}$max, presumably owing to the reduced muscle mass dependent on chronic hypoxia.

To sum up, in acute hypoxia, maximal aerobic performance is primarily reduced by the drop in Hb oxygen saturation, whereas the factor limiting aerobic exercise in chronic hypoxia is mainly located in the muscles; the drop in the body's bicarbonate stores (i.e. the reduction in the body buffer capacity) is associated to some extent with a reduction in the maximal anaerobic (lactic) capacity of both natives and acclimatized lowlanders; finally, prolonged exposure to hypoxia (~ five weeks) produces a marked decrease in peak anaerobic power.

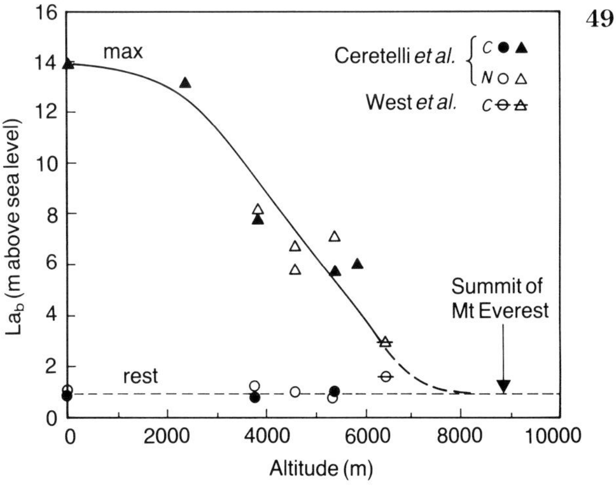

49

49 Resting and maximum blood lactate concentration (La_b) at altitude.

C = Caucasians ● ▲ ⊖ △.
N = altitude natives (○ = Peruvian Indians; △ = Sherpas)

Training for high-altitude climbing

Oswald Oelz

Climbing in the higher regions of the Alps, the Andes and the Himalayas generally consists of walking and scrambling over glaciers and rugged areas, in conditions of low oxygen pressure. It is therefore the lower extremities of the body and the cardiorespiratory system that are of most importance to high-altitude climbers. Good muscular build-up of the legs, a high $\dot{V}O_2$max and a high anaerobic threshold are clearly desirable. These features may be developed through prolonged aerobic running at around 70% of the maximal heart rate, with occasional thrusts toward or beyond the anaerobic threshold. In contrast, modern rock and ice climbing, which demands strong muscular power of the shoulders and arms, is generally not involved in high- and extreme-altitude ascents. There are several approaches to attaining the standard of physical fitness necessary to succeed on peaks at these altitudes, or during prolonged alpine climbs, and the method adopted depends on the likes and dislikes of the individual climber.

In central Europe, particularly Germany, the serious approach is practiced. This should involve at least three work-outs per week, of 20 or more minutes at 70–80% of maximal aerobic power. Surprisingly, it is mainly the novices in Himalayan climbing who engage themselves in heavy work-outs, in order to attain success on a subsequent first expedition, whereas the veteran Himalayan climber gets into shape on the walk into the mountain and during the climb. Presumably, it is better still to run up hills and mountains during these work-outs, although this entire field of anecdotal experience lacks any substantial data on correlations between training methods and subsequent success at high and extreme altitude.

The serious approach to training for high-altitude climbing, by daily uphill running for over 20 minutes, was used by Messner (**50**) before his historic ascent of Everest without supplementary oxygen. However, owing to tight time schedules, and also to the slowly developed conviction that occasionally life should be enjoyed, Messner later adopted a more casual approach toward training for extreme altitude. With the exception of some walking and a little running, Messner did not train when he completed the series of fourteen 8,000 m peaks. He tried to acclimatize and to gain the benefits of training during the approach march to the mountain, and during the initial route preparation. Thanks to this almost sloppy attitude, he gained weight in the lowlands, and created a welcome reserve for himself higher up the mountain, where availability of and desire for food were rather limited. Generally speaking, this is a reasonable recommendation for other climbers – gaining a few extra kilos of body weight before a major expedition is considered good preparation. History shows it worked for Messner. However, the thin and bony frames of some of the most successful modern Himalayan climbers also illustrate that weight gain is not a mandatory prerequisite.

50

50 R. Messner on the south face of Aconcagua (6,995 m above sea level).

Another more casual approach toward training is practiced by the British climbers and is best illustrated by Doug Scott's statement: 'We don't train for climbing, we climb.'. This is even true of Polish extreme-altitude climbers, who represent the toughest and most successful mountaineering group, in terms of successful ascents of 8,000m peaks, under rough circumstances. Jerzy Kukuczka, who climbed all fourteen 8,000m peaks either by new routes or in winter, never trained. He was even known to smoke heavily in the lowlands, though he stopped as soon as the expedition started. However, the author does not recommend smoking as a training method for Himalayan climbing!

Nutrition at altitude

Paolo Cerretelli

Altitude exposure up to 5,000m is often accompanied by an initial loss of body weight that can be attributed to acute mountain sickness (AMS). In fact, anorexia, nausea and vomiting are frequent occurrences in the early phase of acclimatization, and have negative effects on caloric intake and energy balance. At moderate altitude, once the symptoms of AMS disappear, food intake is usually resumed and body weight can stabilize.

For decades, a high carbohydrate intake has been shown to improve both the physical and mental performance of the mountaineer (Consolazio *et al.*, 1969), and is also recommended to alleviate the symptoms of AMS. At any given altitude, the $Pa\text{O}_2$ and, as a result, its O_2 concentration, is influenced by the type of substrate utilized by the tissues. When exercising at altitude, it may be shown that upon oxidation of glucose (with a respiratory quotient, $RQ = \dot{V}\text{CO}_2/\dot{V}\text{O}_2$, approaching unity), the saturation of hemoglobin with O_2 may increase markedly (at 5,000m, from 80 to 85%). In practice, optimizing glucose oxidation is tantamount to lowering the altitude.

Mountaineering involves prolonged, repeated efforts, causing a total depletion of the body's glycogen stores. At sea level, the replenishment of these stores usually takes 24–48 hours, the rate of glycogen resynthesis being greatly influenced by the diet. A carbohydrate-rich diet is therefore recommended, in order to speed up the recovery process. The latter may be negatively influenced by hypoxia, though direct measurements of the rate of glycogen resynthesis in man at altitude are not available. Anecdotal information from mountain climbers unanimously indicates a serious physical deterioration, evidenced by a remarkable loss of body mass (about 10% of the initial weight after 8–12 weeks' exposure), at altitudes exceeding 5,000m.

Studies carried out during the 1981 American Medical Research Expedition to Everest (AMREE) have provided a clear-cut description of the changes in body composition and intestinal absorption, and of endocrine changes in a large group of its members (Boyer and Blume, 1984).

The observed decrease in body weight was subdivided into two phases: the first, lasting up to arrival at the base camp at 5,400m (20 days' duration), during which the average weight loss (~ 2kg) was mainly attributed to a reduction in body fat stores (possibly due to training); and the second, stretching over the following 45 days (5–8kg), characterized by a further loss of fat, but mainly of the muscle mass. After 8–12 weeks' exposure to altitudes between 5,400 and 8,000m, the average loss of body weight is usually about 10% of the initial weight. Muscle mass, based both on anthropometric measurements (using computerized tomography of limbs), and on the determination of the diameter of muscle fibers from biopsies taken from the vastus lateralis muscle, undergoes a 10–15% reduction. This reduction is accompanied by an even more pronounced decrease of the mitochondrial volume (Hoppeler *et al.*, 1990). The loss of muscle mass, together with the reduction in the activity of some of the enzymes of the Krebs cycle, is the main determinant of the decrease in the maximal aerobic and anaerobic power of acclimatized lowlanders.

A three-day study, carried out by Boyer and Blume (1984) in some members of the AMREE, showed a 25% reduction in overall caloric intake at 6,300 m. Carbohydrates represented 63% of the total, vs. 54% at sea level. Fat absorption, which at sea level is normally around 100%, fell by about 50%. In addition, the absorption of xylose was significantly reduced. The caloric deficit found was therefore attributed both to reduced caloric intake, and to intestinal malabsorption. Evidently, the body initially mobilizes the fat from the stores, and subsequently the proteins from the muscles. From recent unpublished experiments, it appears that the reabsorption of amino acids in chronic hypoxia is normal and therefore cannot be a determinant of the reduction in muscle mass.

At an altitude of 6,300 m, in the fasted metabolic state, Boyer and Blume (1984) also provided evidence of an increased fat and protein turnover. They found an increase in the circulating level of triglycerides, as well as of total proteins and albumin. In addition, glucose loading failed to produce the serum glucose increase found at sea level, and the increase in serum insulin following glucose loading was less pronounced. According to the authors, these data are compatible either with an impaired rate of intestinal glucose reabsorption, or with an enhancement of the tissue sensitivity to insulin.

Together with increasing the spontaneous caloric intake, which often appears to be insufficient, the acclimatized mountaineer should try to minimize the loss of proteins. However, it is not clear at this time whether a supplement of proteins in the diet might be beneficial.

The circulating level of a number of hormones influencing metabolism has also been determined, after prolonged high-altitude exposure in the course of the AMREE. The main observed changes were increases in triiodothyronine (T_3), thyroxine (T_4) and norepinephrine. As is well known, T_3 and T_4 increase cellular oxidation of glucose and mobilize fat, whereas norepinephrine increases the glycogen turnover in the liver and gluconeogenesis. These hormonal changes are consistent with the metabolic adaptations described above.

Sleep at altitude

Paolo Cerretelli

Insomnia is a common occurrence at high altitude, and is often accompanied by headache. Both may be minor symptoms of acute mountain sickness (AMS). Recently, some light has been shed on the nature and mechanisms of sleep disturbance, on its relation to AMS, and on the effect of sedatives.

Sleep disturbance at altitude consists of a deterioration in the quality of sleep. The duration of light sleep (stages I and II) increases at the expense of deep (stages III and IV) and rapid eye movement (REM) sleep. In addition, the number of awakenings increases, the arousals being often (though not exclusively) associated with the increased phase of periodic breathing. The latter phenomenon, first described by Mosso in 1898, is particularly important at altitudes exceeding 6,000 m, because of the very low levels of arterial $P\text{O}_2$ that may develop following apneic periods. The low $P\text{O}_2$ leads to hyperventilation, which, in turn, is linked to the arousal. In **51**, the phasing of arterial $P\text{O}_2$ saturation and heart rate relative to tidal volume is given for a member of the 1981 American Medical Research Expedition to

Everest, at 6,300m (West *et al.*, 1986).

The administration of O_2 during sleep at altitude has proven beneficial, both in terminating periodic breathing and in improving sleep.

The effects of acclimatization on altitude insomnia have been less well studied. Natani *et al.* (1970) indicate continuous reductions in stage IV and REM sleep. Other factors besides those described above (e.g. psychological) could affect sleep during prolonged high-altitude exposures. On the other hand, acclimatization may produce an 'altitude lowering effect' on sleep hypoxemia, and thus an improvement in sleep.

The use of sleeping pills at altitude, particularly of diazepam, is probably 'an unwise choice' (Powles, 1982), as it increases the duration of stage II sleep at the expense of stage IV sleep. Thus, one sleeps more, but less deeply. In addition, it has been found that after taking 10mg of diazepam, hemoglobin oxygen saturation decreases significantly.

In order to prevent altitude insomnia with its negative effects on physical performance and judgement, some authors (e.g. Sutton *et al.*, 1979) have attempted to aid acclimatization and sleep with acetazolamide. The results appear to be positive, both with respect to arterial oxygen saturation, which is unaffected by sleep, and with respect to the number of arousals. However, as this drug causes acidosis, it may impair physical performance.

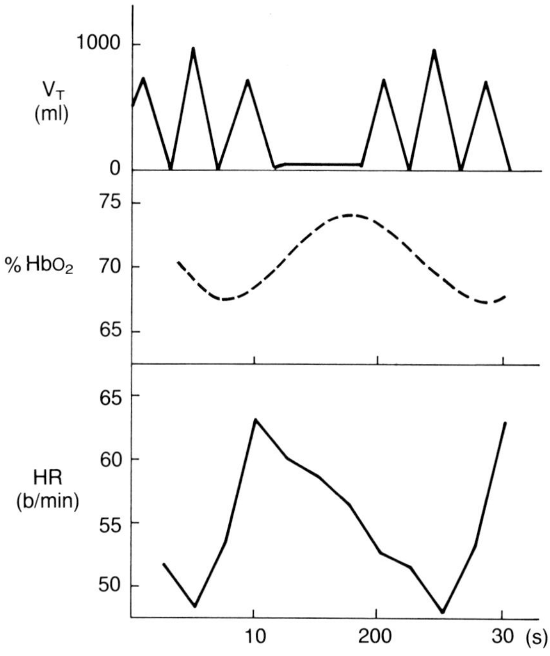

51 Phasing of arterial Po_2 saturation (% HbO_2) and heart rate (HR), in relation to tidal volume (V_T). *Redrawn from West et al., 1986.*

Acute mountain sickness

Oswald Oelz

Acute mountain sickness (AMS) is a syndrome whose early symptoms include headache, anorexia, lassitude and insomnia, possibly progressing to vomiting, ataxia and pulmonary as well as cerebral edema. It occurs in unacclimatized individuals who rapidly ascend to altitudes of 2,500–3,000m or more, and takes from a few hours to a few days to develop. AMS is caused by the direct and indirect effects of hypoxia, which are still poorly understood. Spasms or dilatation of cerebral blood vessels due to hypocapnia, which primarily causes vasoconstriction and cerebral vasodilatation as a result of hypoxia, may interfere during the development of cerebral forms of AMS. Furthermore, hypoxia and exercise stimulate the renin–aldosterone system, favouring the retention of salt and fluid. Hypoxic pulmonary hypertension, causing dilatation of the right heart, seems to stimulate the secretion of the atrial natriuretic peptide, which may also contribute to edema formation through its effects on vascular permeability.

The likelihood of AMS developing in a given individual depends on that person's disposition, the speed of ascent to altitude, and the amount of exercise in the hypoxic environment. A climber with a history of AMS is more likely to develop AMS-related symptoms during altitude exposure. However, this susceptibility can vary from time to time. The most likely factor contributing to individual susceptibility to the development of AMS is the hypoxic ventilatory response. This determines the degree of alveolar hyperventilation that compensates for hypoxia. The speed of ascent is crucial: in a field study in Nepal, Hackett and his coworkers (1976) found that people who flew to an altitude of 2,800m, and subsequently walked to 4,200m, had an AMS incidence of 49%, as compared to 27% among those who walked the entire way. Individuals who ascend in the Alps to 4,550m within 24–48 hours and stay there, have an AMS incidence of over 50%. Where the ascent is strenuous, this incidence may be even higher, as exercise leads to more severe oxygen desaturation of the blood.

Mild AMS, or what Dr Hackett calls 'feeling the altitude' (1976), is characterized by a few mild symptoms like headache that is responsive to mild analgesics, anorexia, and a degree of lassitude and insomnia. Affected individuals also have a higher incidence of periorbital and, eventually, other peripheral edemas. Dyspnea on exertion, palpitation, and feeling hot due to an elevated body temperature are also frequently reported. These symptoms are usually self-limited and disappear within 24–48 hours, if the affected individual abstains from further vigorous ascent and exercise. However, if a rapid ascent to altitudes of 4,000m or higher has been performed, or if symptoms are neglected, more serious symptoms may develop: vomiting, severe headache not relieved by analgesics, dizziness, ataxia, tachypnea and pulmonary rales are the hallmarks of severe AMS, and should be taken as warnings that pulmonary and cerebral edema are imminent. Although even these symptoms may be self-limited, an immediate descent to lower altitudes and, eventually, other measures also are mandatory.

The treatment of mild AMS consists mainly of rest days at the same altitude, appropriate fluid intake to keep the urine clear, and, eventually, aspirin or related drugs for the treatment of headache. Most cases of mild AMS will respond to these measures, and the affected climber can subsequently ascend to higher altitudes at a somewhat slower pace. The administration of acetazolamide in a dose of 250–500mg per day generally facilitates this recovery. Acetazolamide may be particularly helpful in cases of mild AMS associated with fluid retention, weight gain and marked periorbital and peripheral edemas. Dexamethasone may also improve recovery under these conditions, but should be reserved for more serious cases of AMS.

The treatment for serious AMS with excruciating headaches, ataxia and/or tachypnea, is descent or evacuation to a lower altitude, or, if this is impossible, treatment with oxygen. Occasionally, such treatment is prevented by bad weather or avalanche conditions, and oxygen equipment is not available. In these circumstances, dexamethasone in an initial dose of 8mg, followed by 4mg every six hours should be given (Ferrazini *et al.*, 1987). This treatment has been found to reduce the severity of AMS markedly, particularly in patients with cerebral AMS symptoms. However, a patient may improve with regard to cerebral symptoms, but nevertheless develop high-altitude pulmonary edema (HAPE) while under treatment with dexamethasone.

AMS is prevented in most cases by slow ascent, facilitating acclimatization. 'Slow ascent' is hard to define: the speed of ascent that is perfectly tolerable for one individual may cause HAPE in another. Therefore, each climber should pay attention to the warning signs of his body and respect them. Most climbers are safe if they follow this rule of thumb: above 2,500–3,000m, the sleeping altitude should not be increased by more than 300m per 24 hours. Occasional rest days should be taken when needed by the climber, and days of heavy exertion should also be followed by rest days. Adequate fluid intake must be maintained, resulting in the production of at least 1 liter of clear urine per 24 hours.

Climbers who become sick in spite of these preventive measures should take acetazolamide for prophylaxis of AMS. Acetazolamide is a carbonic anhydrase inhibitor, and produces a metabolic acidosis by increased urinary excretion of bicarbonate. As a result, respiration is stimulated, and the periodic breathing during sleep at high altitude which causes severe oxygen desaturation is reduced. The side effects of acetozolamide, such as tingling, alteration in the taste of carbonated drinks and increased diuresis, stop when the drug is discontinued. In several well-controlled studies, acetazolamide has been shown to reduce the severity of AMS and to improve oxygenation and the performance of climbers at high altitude. Although various doses have been used, a daily morning dose of 250 or 500 mg in the form of a slow release preparation is presently recommended by most experts.

Dexamethasone has also proven useful for the prophylaxis of AMS (**52**). However, the long term side effects of corticosteroids make this sort of prophylaxis less desirable. Other diuretics, such as frusemide, are dangerous and should definitely not be given for the prevention of AMS.

52

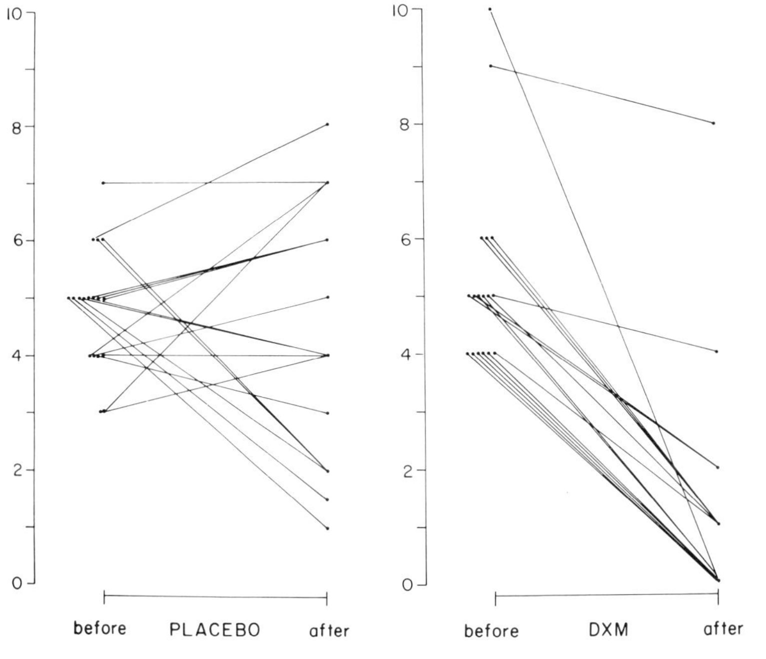

52 The acute mountain sickness score of mountaineers before and after 12–16 hours of treatment with a placebo or dexamethasone.

Acute mountain sickness: risk factors

Jean-Paul Richalet

A few observations have shown that it is nearly always the same subjects who are likely to adapt to high altitude without difficulty. In other words, acute mountain sickness (AMS) occurs more frequently in people who have already suffered from this syndrome. These subjective observations may lead to the conclusion that there exist some risk factors which predispose a subject to AMS. Some authors have shown that AMS, pulmonary edema, and physical performance in high altitude could be related to the hypoxic ventilatory drive (Hu *et al.*, 1982; King and Robinson, 1972; Masuyama *et al.*, 1986; Mathew *et al.*, 1983; Moore *et al.*, 1986; Schoene *et al.*, 1984). Recently, the intensity of hypoxic ventilatory drive has been directly linked with the nervous response of chemoreceptors (Vizek *et al.*, 1987). Thus, it might be suggested that people who suffer quite severely from AMS are those who, genetically, have a poor chemoreceptor sensitivity to O_2 deficiency in the blood.

In a three-year study developed in the Faculty of Medicine at Créteil, the risk factors of AMS in 128 high-altitude climbers (26 females and 102 males) were systematically scanned. Subjects were tested prior to their departure on an expedition to high altitude (6,200–8,848 m), and clinical signs of AMS were recorded during the expedition (**53**). A discriminant analysis was performed to assess the risk factors of AMS. Aerobic power was assessed by direct measurement of $\dot{V}O_2$max. Ventilatory and cardiac responses to hypoxia (11.5% O_2) were determined both at rest and during exercise (50% $\dot{V}O_2$max). Autonomic nervous response was evaluated by means of a cold pressor test (CPT) (Viswanathan *et al.*, 1978). Visual field and eye fundus were examined before and after the expedition. Some of the parameters obtained during this study are shown in **Table 1**. For more detailed results, see Richalet *et al.* (1988).

53

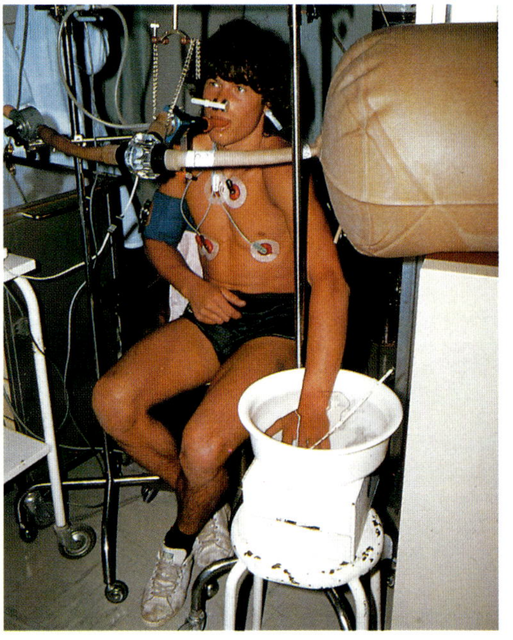

53 Hypoxic tests.

During the expeditions, three climbers died (two men and a woman), probably from altitude-induced physiological or psychological degradation. 37% of the men and 65% of the women suffered from severe AMS. Risk factors of AMS appeared to be: previous history of severe AMS, or headache at sea level; low ventilatory and cardiac response to hypoxia; rapid and superficial pattern of ventilation; blunted response to CPT. Endurance training and regular climbing in the Alps do not protect against AMS. Atopy and smoking do not favor its occurrence. Physical performance, evaluated by the maximal altitude reached (6,800 ± 1,100 m), is closely related to sea level $\dot{V}O_2$max (50.7 ± 7.1 ml.min^{-1}.kg^{-1}) (**54**), but not to hypoxic responses. The elite climbers (n = 22) had a $\dot{V}O_2$max of 55.8 ± 6.4 ml.min^{-1}.kg^{-1}, lower than elite, endurance-

Table 1

Men	*AMS* −		*AMS* +	*Elite* −		*Elite* +
n	64		38	80		22
age	35 ± 7		36 ± 8	36 ± 8	*	32 ± 7
Max alt.e. (km)	6.8 ± 1.1		6.8 ± 1.0	6.5 ± 0.9	***	8.0 ± 0.6
Success (%)	86 ± 12		87 ± 11	84 ± 11	***	97 ± 6
$\dot{V}O_2max$ ($ml.min^{-1}.kg^{-1}$)	50.4 ± 7.4		51.2 ± 6.7	49.3 ± 6.7	***	55.8 ± 6.4
$\Delta HR/\Delta SO_2E$ ($min^{-1}.\%^{-1}$)	−1.07 ± 0.54	*	−0.86 ± 0.33	−0.96 ± 0.46		−1.10 ± 0.54
$\Delta \dot{V}E/\Delta SO_2E$ ($l.min^{-1}.kg^{-1}$)	−1.10 ± 0.73	**	−0.80 ± 0.37	−0.96 ± 0.66		−1.09 ± 0.48
V_T ex.h. (l)	2.7 ± 0.5	***	2.3 ± 0.4	2.6 ± 0.5		2.6 ± 0.5

Table 1. Physiological characteristics of male climbers (n = 102).
AMS + = subjects suffering from severe AMS; AMS − = subjects suffering from light or moderate AMS; Elite + = subjects with a background of high-level climbs in the Himalayas.
Max alt.e. = maximal altitude reached during the expedition; Success = percentage of success in terms of altitude (Maximal altitude reached/Altitude of the summit × 100); $\Delta HR/\Delta SO_2E$, $\Delta \dot{V}E/\Delta SO_2E$ = cardiac and ventilatory responses to hypoxia, when exercising (E); with ΔHR = variation in heart rate between hypoxia and normoxia, $\Delta \dot{V}E$ = variation in total ventilation between hypoxia and normoxia divided by body weight, ΔSO_2 = variation in O_2 arterial saturation between normoxia and hypoxia; V_T ex.h. = tidal volume when exercising in hypoxia; Mean ± S.D., Student test = * $p<0.05$, ** $p<0.01$, *** $p<0.001$.

trained athletes, as already shown by Oelz *et al.* (1986). Retinal hemorrhages (found in 9% of subjects) and muscular mass loss were related to maximal altitude reached.

AMS high risk subjects could be detected by means of discriminant analysis. A history of migraine or headaches at sea level, a pattern of ventilation, and hypoxic drives seem to be the most pertinent factors to look for, if one wants to detect susceptible subjects and warn them to make a particularly slow ascent to high altitude.

From the study, it appeared that the main determining factor for performance at high altitude was $\dot{V}O_2max$, as shown in (Richalet *et al.*, 1987, 1988). Conversely, maximal altitude reached was not related to hypoxic drive.

54 Relation between the maximal altitude reached by the climbers and $\dot{V}O_2max$ determined at sea level before the expedition (men, n = 102, $p < 0.001$).

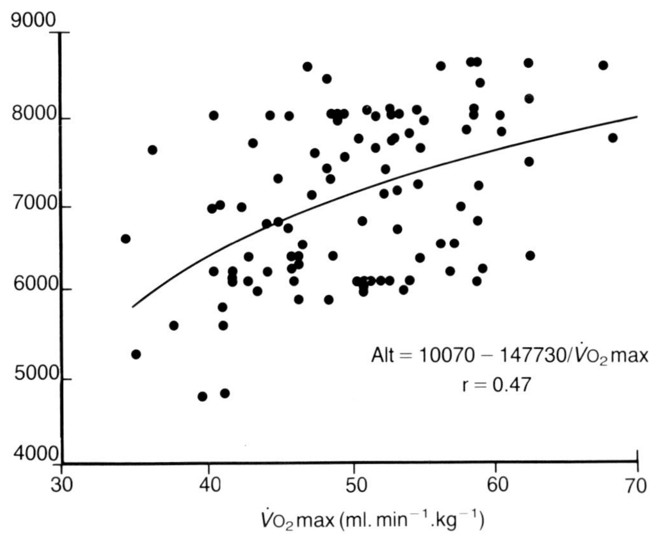

54

These two main results, i.e. that occurrence of AMS is linked to hypoxic drive, and that performance in high altitude is linked to $\dot{V}O_2max$, are not contradictory. Both reflect the time course of acclimatization to high altitude, as shown in **55**. The period of exposure to high altitude may be divided into four phases. **Phase 1** (blank phase): 0–6 hours, when no symptoms of AMS are felt. **Phase 2** (acclimatization phase): 6 hours–7 days, when all the ventilatory, renal, hematopoietic and cellular processes of acclimatization develop. **Phase 3** (phase of completed acclimatization): 7–21 days, when subjects are acclimatized and ready to perform a high level of energy expenditure in high altitude. **Phase 4** (degradation phase): more than 21 days, when the organism begins to suffer from prolonged exposure to very high altitude, resulting in weight loss, muscle mass loss, chronic dehydration, high blood viscosity, etc..

AMS mainly occurs in Phase 2, and high hypoxic drives are very important during this period, when the subject is acutely exposed to hypoxia. Performance is generally at its best in Phase 3, when AMS symptoms have usually subsided, and $\dot{V}O_2max$ – a determining factor of aerobic, long-lasting exercise, such as is performed in high-altitude expeditions – is an important factor for those who want to climb high and quickly. A subject with a good hypoxic drive and a low $\dot{V}O_2max$ will not suffer from AMS, but will not be able to climb very quickly above 7,000 m. Inversely, a subject with a low hypoxic drive and a high $\dot{V}O_2max$ might be very sick during the first days of acclimatization, but if he gets through this period doing little exercise and waiting for a better acclimatization, he can perform very well a few days later at high altitude.

In Phase 2, good acclimatization will result if a steady, progressive gain in altitude is respected (less than 300–500 m of altitude gain per day, averaged over two consecutive days), as the major factor that induces AMS is a quick rate of ascent, whatever the climber's response to hypoxia. However, in order to climb an 8,000 m peak, it is necessary to get acclimatized at a sufficiently high altitude. The optimal altitude for a base camp is 4,800–5,200 m. In Phase 4, the de-

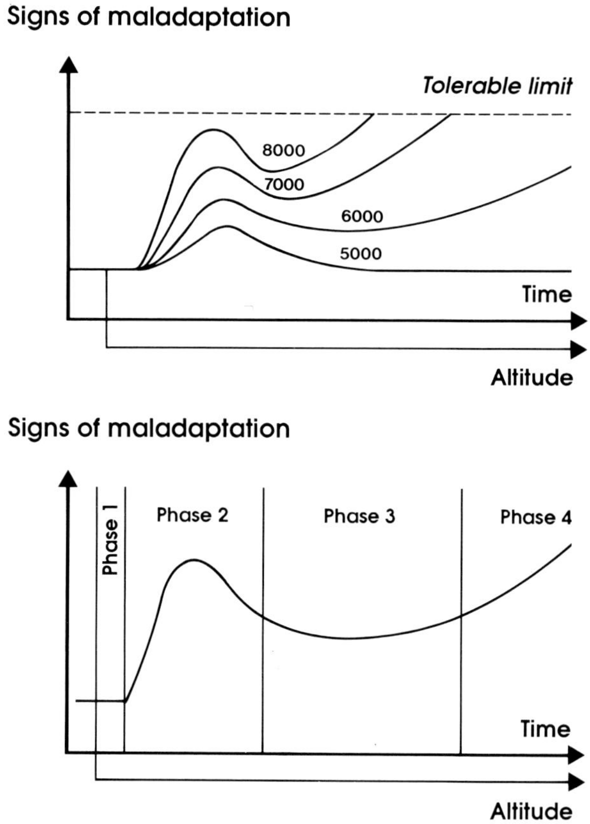

55 The four phases of physiological reaction to exposure to altitude hypoxia.

gradation will be partly avoided if the climber spends less than six nights above 6,000–6,500 m, and no more than four consecutive nights. The following three 'golden rules' summarize these points:

- 'Don't go too high too fast.'
- 'Go high enough to get acclimatized.'
- 'Don't stay too high too long.'

In conclusion, high-altitude climbers do not seem to have more highly developed physiological characteristics than do average, moderately trained athletes. However, a good response to hypoxia protects against severe AMS during the first days of acclimatization, and a good $\dot{V}O_2max$ allows a better and safer performance in very high altitude. Even if you have a good response to hypoxia, you can suffer from AMS if you don't follow the three major rules of progression in high altitude.

Localized or peripheral edema

Oswald Oelz

Swelling of the ankles and feet, the hands and, more commonly, around the eyes and in the face, is common during gradual exposure to altitude. At an altitude of 4,200 m in Nepal, Hackett and Rennie (1979) found that 23% of trekkers had at least one area of peripheral edema. These edemas are found less frequently in the Alps, and are twice as common in women as in men. People with peripheral edema usually experience some degree of weight gain, as the edemas represent sodium and water retention, as well as some degree of hypoxia-induced capillary leakage, and abnormal fluid distribution. Edema of the hands may be caused by rucksack strap compression, movement of the arms and other local factors. Edema of the feet is rare. Periorbital edema is more characteristic of acute mountain sickness (AMS) (**56**).

Although peripheral edema is harmless in the majority of cases, it may be associated with more severe symptoms and signs of AMS. Therefore, people with peripheral edemas should be checked for other manifestations of AMS, and for pulmonary edema and, if these are present, the appropriate measures should be taken. Severe swelling around the eyes may impair vision and, in itself, represent an indication for descent or acetazolamide. Under these circumstances, even frusemide may be indicated. A low salt diet, acetazolamide and gradual ascent is the best prophylaxis against peripheral edema for susceptible persons.

 **56**

56 Periorbital edema.

Polycythemia and thromboembolic episodes

Oswald Oelz

High-altitude hypoxia stimulates erythropoietin, and consequently red cell production. Maximal hematocrit values of 55–60% have repeatedly been measured during expeditions in the Himalayas, even in well hydrated climbers. Higher hematocrit values up to 78% have been reported, and are the result of both polycythemia and dehydration. At these extreme values, blood viscosity increases exponentially and the microcirculation is impaired, favoring the development of frostbite and thrombosis. Activation of intravascular coagulation has been demonstrated in patients with established high-altitude pulmonary edema, but not during the development of this syndrome. It is thus an epiphenomenon of edema formation, and not a pathogenetic factor.

There are numerous reports of thromboembolic episodes in Himalayan climbers presenting mainly as pulmonary embolism, or as neurological symptoms. These range from transient ischemic attacks

(TIA) to strokes. TIAs are reported with increasing frequency by climbers not using supplementary oxygen at extreme altitudes. Several cases of sudden death occurring above 7,000 m in the last few years could probably be explained as thromboembolic episodes. The actual incidence of vascular complications at extreme altitude may be much higher than suggested by the literature on the subject, as many of these events remain undiagnosed, even at sea level. Dickinson (1983) actually observed widespread pulmonary and/or cerebral thrombosis in five out of seven victims of severe acute mountain sickness (AMS) at autopsy in Nepal.

Patients with cerebrovascular problems or symptoms suggestive of pulmonary embolism should descend or be evacuated as fast as possible, and generous fluid replacement should be implemented. The possible benefit of heparin must be weighed against the danger of potential hemorrhagic complications, e.g. from retinal or cerebral bleeding.

Thromboembolic complications can be avoided by ensuring a proper fluid intake, resulting in a daily urine production of at least one liter of clear urine, and by prevention of AMS and high-altitude pulmonary and cerebral edema. When storm-bound in a small tent, climbers should occasionally exercise their legs to avoid deep venous thrombosis. Aspirin (not more than 100 mg/day) may be helpful in preventing thrombosis, but may aggravate retinal bleeding. Climbers who experience cerebrovascular problems at extreme altitude should be strongly discouraged from returning to the same altitude.

High-altitude pulmonary edema

Oswald Oelz

High-altitude pulmonary edema (HAPE) develops in otherwise healthy individuals who rapidly ascend to altitudes of 3,000 m or higher. Occasionally, HAPE has been observed at even lower altitudes. The symptoms of HAPE may appear suddenly in a healthy climber, within 24–72 hours of high-altitude exposure or, more commonly, develop gradually in a person already suffering from acute mountain sickness (AMS). Contributing factors are strenuous exercise and cold temperatures.

People developing HAPE suffer initially from a non-productive cough, shortness of breath, weakness, fatigue and other AMS symptoms. These symptoms are exacerbated while in the supine position, and tachypnea develops. Later, the cough becomes productive with watery, and sometimes bloody, sputum. Rhonchi and gurgling sounds can eventually be heard without a stethoscope. However, the common clinical signs are crepitant rales, rhonchi, tachypnea, cyanosis and tachycardia. Later on, patients become irrational and hallucinatory due to hypoxia and/or accompanying cerebral edema. This is rapidly followed by coma and death.

57

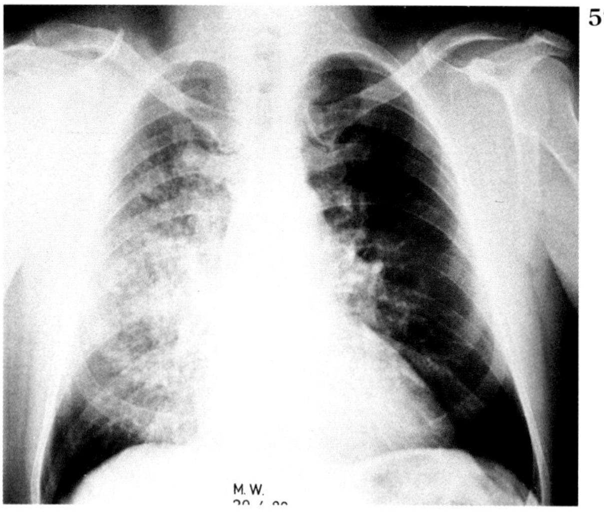

57 Typical chest X ray of HAPE at Capanna Margherita (4,554 m).

The incidence of HAPE in trekkers in Nepal was reported to be 2.5–4.5% in the late seventies, and has since decreased owing to educational measures and awareness of the disease. It affects 1–2% of climbers on Mt. McKinley and is rarer in the Alps. In Switzerland, approximately 10 climbers per year have to be air-rescued due to HAPE.

Laboratory tests on victims of HAPE show a moderately elevated hematocrit and some degree of leucocytosis. The arterial PO_2 is in the range of 2.7–4.0 kPa, and the PCO_2 is usually in the normal range. The arterial oxygen saturation may drop as low as 40%. 100% oxygen therapy does not correct the hypoxemia entirely, pointing to a persisting diffusion limitation and/or pulmonary shunting.

The X ray characteristically shows patchy, not confluent, infiltrates, which are often more severe on the right side (**57** & **58**). One-sided HAPE is observed in 20–30% of cases. The pulmonary arteries appear prominent and the size of the heart is usually normal.

Catheter studies show marked pulmonary hypertension and normal wedge pressure. Normal left ventricular function has also been demonstrated by echocardiography in patients with full-blown HAPE.

The pathophysiological sequence of HAPE is thought to start with alveolar hypoventilation, due to a low hypoxic ventilatory drive. Subsequently, accentuated hypoxic pulmonary hypertension and a capillary leak syndrome lead to a high pressure permeability edema.

HAPE must be treated with immediate descent or evacuation to lower altitudes. This causal treatment should be performed immediately, without spending time on other maneuvers. Positive end expiratory pressure breathing has been reported to improve oxygenation to some extent, and to improve the clinical condition of a few patients. If available, oxygen should be given with a flow rate of 2–4 liter/min. Initial administration of nifedipine (10–20 mg) sublingually, followed by 20 mg of slow-release preparation, repeated every 6 hours, results in clinical improvement, better oxygenation, reduction of alveolar arterial oxygen gradient and pulmonary artery pressure, and progressive clearing of alveolar edema. Although this represents a successful pharmacological emergency treatment for HAPE (Oelz *et al.*, 1989), the treatment of choice remains descent to a lower altitude. When this is impossible and when supplementary oxygen is not available, nifedipine offers the greatest chance of improvement. Such improvement, however, should be exploited to enable descent, not to allow activities to continue at high altitude.

58

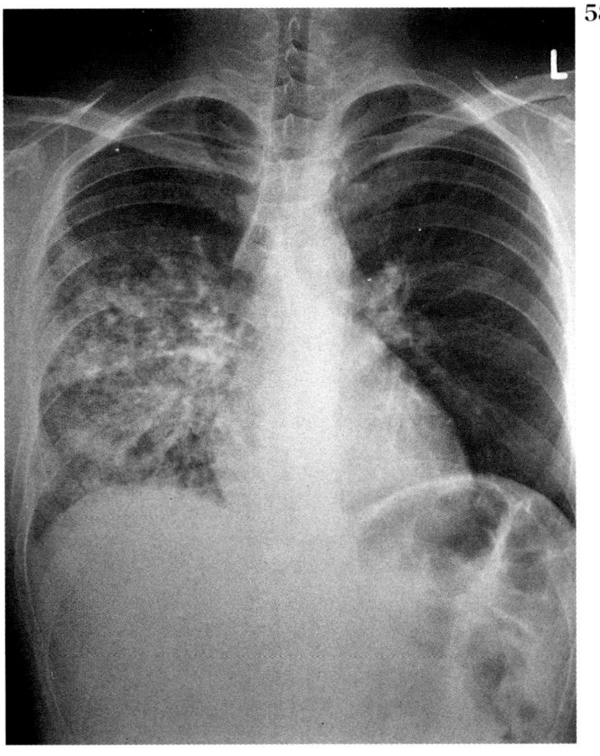

58 Chest X ray of HAPE on Aconcagua.

High-altitude cerebral edema

Charles Clarke

High-altitude cerebral edema (HACE) is one of the conditions caused by chronic hypoxia, which may occur at any altitude above 3,000m. Acute mountain sickness (AMS), peripheral edema, retinal hemorrhage, subcutaneous hemorrhage, high-altitude pulmonary edema (HAPE) and cerebral edema are all related conditions, and may frequently coexist.

Cerebral edema of acclimatization

Cerebral edema of acclimatization is a common condition, reported in around 1–2% of those who sleep above 4,000m before being well acclimatized. Mild cerebral edema blends into AMS. The patient complains of headache, which is often worse on waking, and there is a postural component, the pain being brought on or exacerbated by lying down or straining. The earliest sign is a vague unsteadiness of gait. Papilledema may be present, but is not invariable. The symptoms may resolve at this stage or, if ascent continues, may progress to produce a variety of neurological signs: disorientation, somnolence progressing to coma or dysarthria, nystagmus and hemiparesis may be seen. If the patient loses consciousness, the outlook is grave.

As a general rule, ataxia of gait is the single most useful clinical feature; if a sick climber is unsteady on his feet at altitude, HACE must be assumed, even in the absence of any other sign.

Cerebral edema of extreme altitude

HACE is *usually* a problem of acclimatization, occurring during the first few days at altitude. However, it may occur at extreme altitudes (above 7,000m), sometimes with great rapidity (**59**). Many of the cases are fatal.

59

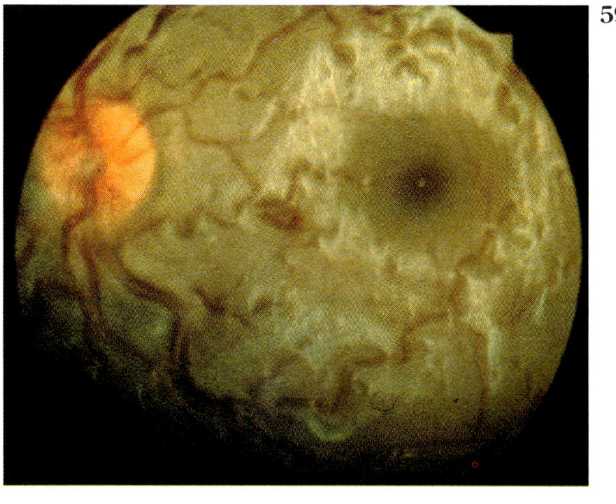

59 Severe papilledema in a Sherpa at 7,000m.

Neuropathology

In HACE, there is widespread cerebral edema with flattening of the sulci, and herniation of both the unci of the temporal lobes and the cerebellar tonsils may be seen. There are many changes in the blood vessels: microinfarction with ring hemorrhages (**60**), massive brainstem infarction, subarachnoid hemorrhage and dural venous sinus thrombosis (**61** & **62**).

60

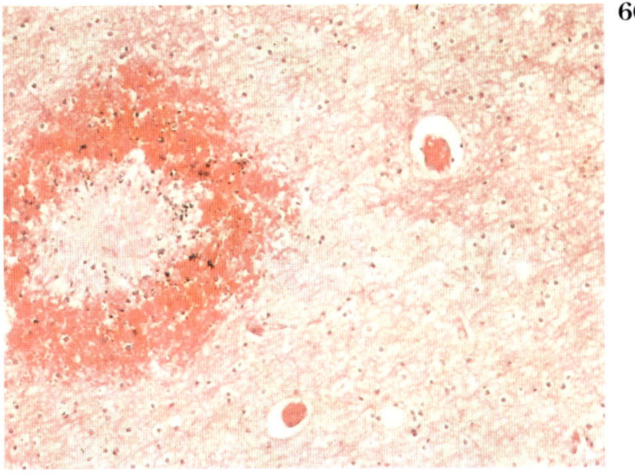

60 Brain edema and ring hemorrhage. *Courtesy of D. Heath.*

61

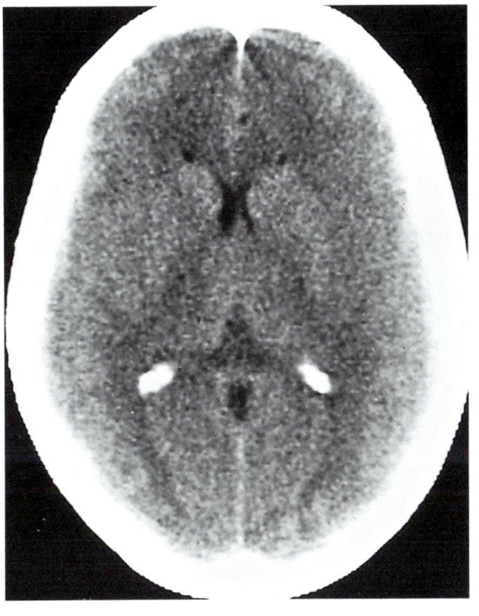

62

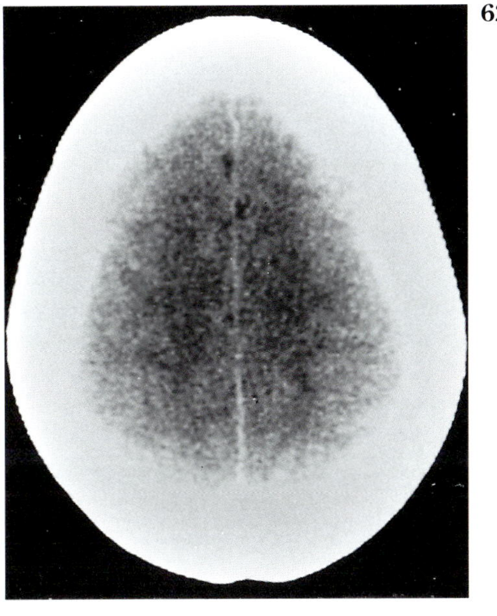

61 & 62 CT scan of cerebral edema, showing effaced cerebral sulci, compressed ventricles and basilar cisterns. Also, the normal contrast between white and grey matter disappears. *Courtesy of R. Meuli.*

Clinical management

There is little doubt that some of the fatalities caused by HACE can be avoided, if the early symptoms and signs are recognized. Ataxia of gait is the single most common sign and should be taken seriously.

Acetazolamide, undoubtedly of value in the

prevention of AMS, may help prevent HACE, but further information is awaited before a definitive statement can be made. Prompt descent is the most important treatment for any form of altitude sickness. Sometimes rapid recovery occurs after a descent of as little as several hundred meters.

Although there are no controlled studies, dexamethasone is almost certainly helpful, and the drug should be given either orally or parenterally in the usual adult doses of 8–16mg per day. Oxygen should be given by mask, if it is available. Compression chambers are also of value.

High-altitude retinopathy

Charles Clarke

At altitudes over 3,000m, the retinal blood flow of a visitor experiences a sharp increase of about 25% of its sea level value, for several days. This increase, together with an increase in the permeability of the retinal capillaries and arterioles, leads to changes in the retina which can be seen with an ophthalmoscope.

Changes in caliber of retinal vessels

The earliest sign of retinal pathology is an increase in the caliber of the retinal vessels. Venous distension is seen, and the veins vary in caliber, appearing 'sausage-shaped' (**63**). These changes will be seen in the majority of climbers at 4,000m. There are no symptoms.

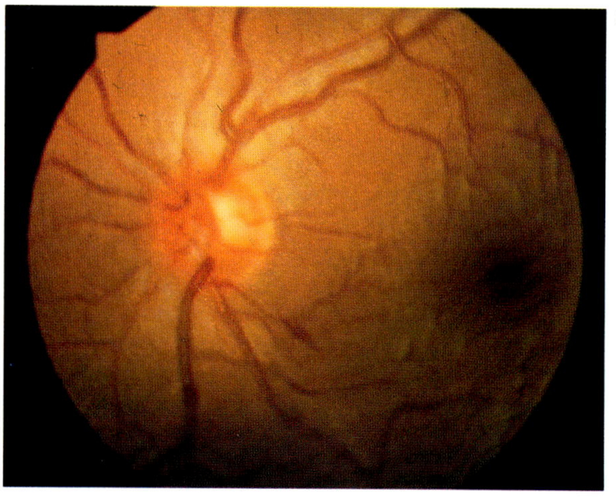

63

63 Venous distension in a symptomless climber at 5,500 m. A small flame hemorrhage is also visible.

Symptomless retinal hemorrhage

Careful studies of the members of large expeditions to high altitude have shown that symptomless retinal hemorrhages occur in about 50% of climbers at 5,000m, during the acclimatization process. The lesions vary from small 'flame' hemorrhages (**63** & **64**) in the nerve fiber layer of the retina, to larger bleeds which are subhyaloid (**65**). These hemorrhages resolve after several days. No treatment is necessary. Exceptionally, a hemorrhage may involve the macula, causing loss of central vision. Recovery is usual.

64 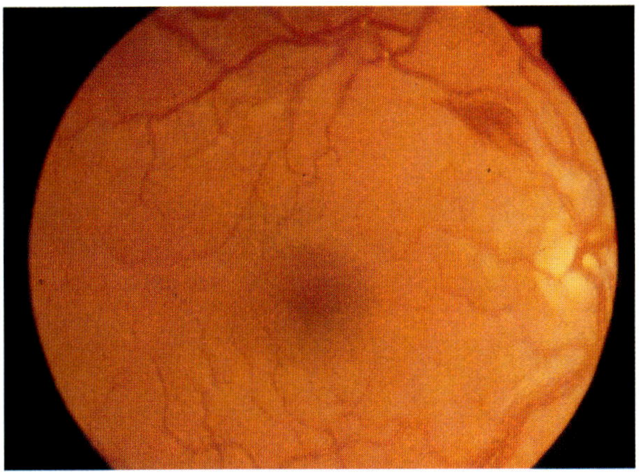

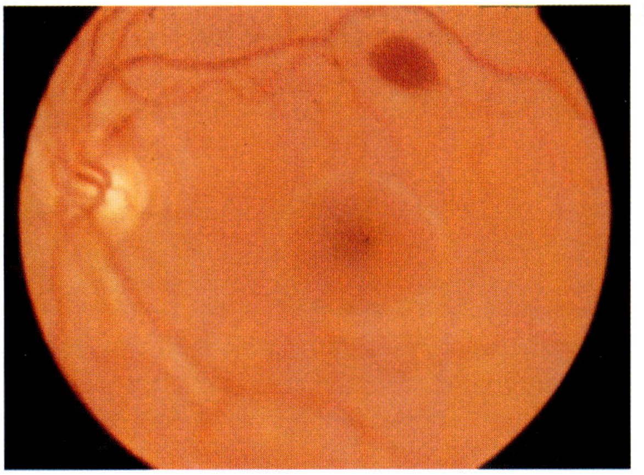 65

64 Flame hemorrhage in a symptomless climber at 6,000 m.

65 Subhyaloid hemorrhage in a symptomless climber at 5,500 m.

Severe high-altitude retinopathy

A severe hemorrhagic retinopathy may occur at altitudes of 5,500–6,000 m or higher, either as an apparent result of insufficient acclimatization, or in climbers who seem well acclimatized (*see also* p. 60). Multiple hemorrhages are present (**66**) and there is severe visual loss which may be permanent.

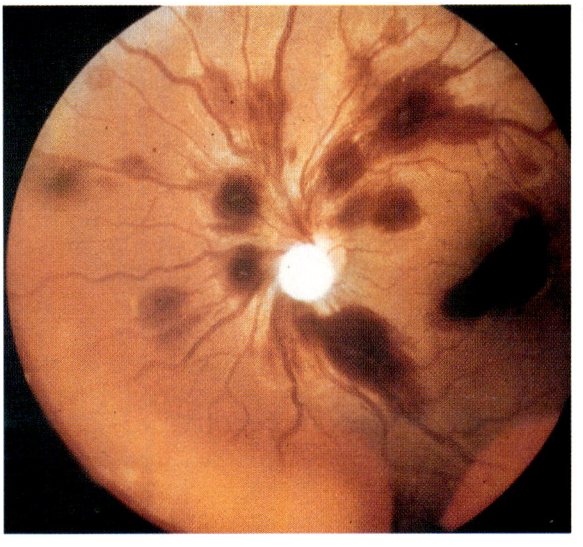 66

66 Multiple hemorrhages in a sick climber at 6,000 m. Visual impairment was permanent. *Courtesy of G. Gobbato.*

Clinical management

Small symptomless hemorrhages require no treatment. Impairment of central vision should be taken seriously and descent advised, preferably below 3,500 m. Dexamethasone is almost certainly of value and, unless contraindicated, should be given as for cerebral edema, 8–16 mg per day.

Mental and neurological disturbances at high altitude

Zdzislaw Ryn

The central nervous system is most sensitive to hypoxia, as a nerve cell uses 20 times more oxygen than a muscle cell. In the first phase of exposure to high altitude, as the body adapts, changes occur in the circulatory and respiratory systems to ensure the nutrition and oxygenation of the brain. The brain itself also displays certain adjustment abilities to high-altitude hypoxia. Among the permanent inhabitants of mountains, a bigger diameter of cerebral capillary, as well as a denser cerebral capillary network, has been observed (Ramos *et al.*, 1967). In addition, a reduced sensitivity to hypoxia, as well as a greater tolerance of anoxia, has been confirmed (Velázquez, 1959). A marked prevalence of voltage in the autonomic parasympathetic system (Monge and Pesce, 1935) is further proof of the adjustment ability of people inhabiting mountainous areas. Cerebral blood flow increases in the initial stages of a stay at high altitude and then, as the body becomes acclimatized, it diminishes (Barragán and Arce, 1988). The pressure of the cerebrospinal fluid increases, and with this comes a rise in the level of glucose, proteins and potassium ions (Severinghaus, 1963). The bioelectrical activity of the brain also changes. In the initial period, an activation and gradual disorganization of basic functions is observed (Querol, 1965; Ryn, 1979). With gradual acclimatization, the voltage of the basic functions decreases and the delta waves disappear (Barragán and Arce, 1988). The speed of impulse transference in the peripheral nerves also decreases.

Psychopathology of acute mountain sickness

In acute mountain sickness (AMS), the type and the intensity of the mental disturbance depends on the altitude, length of stay, and individual psychological predispositions. At low altitudes, up to 4,000 m above sea level, one observes various forms of neurasthenia in most people (80%). This may be of two kinds: emotional excitation or indifference. The above reactions occur during the first stages of exposure to this altitude, and last a few hours or a few days. They disappear spontaneously.

At middle altitudes of 4,000–6,000 m, mental reactions are cyclothymic in character: apathetic–depressive (75%) or euphoric–impulsive (20%). The apathetic–depressive syndrome is characterized by increasing mental and physical fatigue, indifference and loss of interest, aversion to any sort of activity, pensiveness, depressed mood and sadness, drowsiness and diminished sexual interest. Several persons exhibit dysphoria. The euphoric–impulsive syndrome is characterized by an exalted mood, sometimes tinged with euphoria, bewilderment, an irrational feeling of happiness, increased psychomotor drive, unnecessary activity and higher emotional tension with irritability, outbursts, facile interpersonal conflicts and episodes of anxiety.

At high altitudes, over 7,000 m, an acute organic brain syndrome is distinguished. The most frequent symptoms are diminished physical activity, clumsiness, disorientation in time and space, diminished critical ability, labile moods, disorders of memory (amnesia), somnolence, lack of visual–motor coordination and disorders in equilibrium and in perception. In the intellectual sphere, slow, imprecise thinking, dullness, impaired abstract thinking and a tendency to reach false conclusions are noted. Some people at extreme altitude display short-term psychotic symptoms in the form of auditory and/or visual illusions and hallucinations, as well as persecutory or 'wishful thinking' delusions. Some alpinists have described visual hallucinations in the form of coloured objects and shining points, as well as acoustic hallucinations in the form of human voices, sounds coming from great distances, etc. Another characteristic symptom can be excessive drowsiness, a tendency towards suddenly falling asleep for a short time, even when walking. Movement becomes automatic. One alpinist described his condition while on a mountain summit in the following way: 'The problem of life and death was no longer important there, the self-defence mechanism was completely subdued. Had I been allowed to remain there in my sleeping bag, I would have done it, regardless of the consequences. I experienced a complete loss of will and desire to act. I even felt an aversion towards life.'

In rare cases, possible symptoms of irrationality and depersonalization have been observed, as well as disturbances in consciousness, even including coma.

Psychopathology of chronic mountain sickness

Chronic mountain sickness (CMS) is the loss of natural acclimatization. In such cases, the number of red blood cells and the hematocrit increase. High blood pressure in the pulmonary circulation and hypoxia are also observed. The psychopathological picture testifies to limited brain damage. Neurological symptoms are also frequently observed; these are transient and scattered, and generally include numbing, formication, tremors or convulsions, impaired vision, paresis, temporary limb paralysis, and impaired speech similar to aphasia. The cause of these disturbances is most probably embolic, or they may be the result of thrombosis of brain vessels, leading to local or general hypoxia (Monge and Pesce, 1935; Ryn, 1989).

High-altitude cerebral edema

The most dangerous complications of AMS are high-altitude pulmonary edema (HAPE) and/or high-altitude cerebral edema (HACE). The psychopathological pattern of HACE comprises disturbances of the psychomotor drive, the emotional sphere, the memory and orientation, and consciousness with somnolence, stupor and coma (Hajdukiewicz and Ryn, 1983). In severe cases, there are symptoms of generalized and/or local brain injury, with paresis, paralysis, convulsions and consciousness disturbances, including coma. According to Clarke (1988), HACE occurs in two clinical forms: HACE of acclimatization and HACE of extreme altitude. The development of this disease is usually acute and mortality is high. In serious cases, residual cerebral damage may remain for a long time or even permanently. Extreme physical exercise at high altitudes, hypoxia and hypothermia seem to play a part in the mechanism of this disease.

High-altitude cerebral asthenia

Some alpinists affected by the stress of extreme altitudes may even develop permanent damage to the central nervous system. According to the author's studies (Ryn, 1987), high-altitude cerebral asthenia (HACA) occurred in 25% of alpinists who spent a long time at extreme altitudes without using oxygen. The following factors are essential in the development of HACA: the duration of stay at high altitude and the length of time the subject has practiced alpinism, acute organic brain syndrome and Cheyne–Stokes breathing in the course of AMS, and altitude deterioration. HACA can occur in three forms: characteropathic, encephalopathic and neuroplegic. In the first, emotional disturbances prevail; in the second, the predominant symptoms are those of focal brain damage; whereas in the third type, the prevailing symptoms are those of peripheral neurological dysfunction in the form of paresis. This syndrome is chronic and in some cases progressive. Its organic nature is confirmed by the results of physiological and electroencephalographic studies. The intensity of HACA symptoms is usually on a subclinical level. There is no experience in the treatment of HACA. The prophylaxis should comprise adequate preparation before the expedition, competent methods of acclimatization, and detailed medical examinations both before and after each expedition into high mountains.

In some cases, climbing to extreme altitudes can provoke brain damage. The possibility of permanent brain damage fully confirms the opinion of Jack Longland, that high altitudes slowly kill a man.

Solar radiation

The effects of solar radiation

Peter Lomax, Richard Thinney and Bartly J. Mondino

The light emitted by the sun includes radiation of wavelengths between 290 and 1,850 nanometers (nm). The characteristics and proportions of the radiation reaching the earth's surface show considerable variation with the seasons and with changing atmospheric conditions. Radiation of less than 400 nm can cause damage to the skin or to the eyes. Ordinary window glass is able to filter out most of these potentially damaging rays; smoke and smog in urban environments have a similar effect and absorb the majority of the offending rays, especially during the winter months (although a surprising amount of harmful radiation may filter through thin overhead clouds). At high altitudes, these ultraviolet screens are absent and, furthermore, reflection from surrounding rock, sand, snow, ice or water can result in a pronounced increase in the effective exposure (**67**).

67

67 Solar reflexion on a snow surface. *Courtesy of J. Michelet.*

Photophthalmia

At high altitudes, solar radiation can lead to a debilitating, acute, inflammatory reaction of the superficial eye (the cornea and conjunctiva), known as photophthalmia or snow blindness. The latter, more frequently used term is technically a misnomer, as snow is related only indirectly to the disorder, and the associated visual impairment is minor and temporary. Solar radiation in the mountain environment differs from that in other sunny locales, in that the atmosphere is thinner, generally clearer, and allows transmission of more ultraviolet rays. Ultraviolet radiation may rise to 5 or 6% of the total solar radiation at high altitudes, compared with 1–2% at sea level (Duke-Elder *et al.*, 1972; Records and Brown, 1986). Ordinarily, direct exposure to sunlight does not lead to photophthalmia, but exposure for several hours in an environment which adds an element of surface reflection, may damage the eye. A snow or ice field may reflect up to 85% of the incident light waves, while sand or a farmed field reflect only about 17% and 2.5%, respectively (Records and Brown, 1986). When bright, reflected light strikes from below, the eyes are not protected by the eyebrows, forehead or headgear, and this contributes greatly to the cumulative exposure (**68**). The cornea and conjunctiva are particularly vulnerable to damaging radiation, because they lack the melanin and keratinized epidermis which protect the skin.

The ultraviolet spectrum is that portion of the invisible radiation of the sun with wavelengths between 200 and 400 nm. The cornea is opaque to wavelengths of less than 295 nm; rays are either absorbed by the cornea or reflected from its surface. Wavelengths with a peak near 288 nm are able to induce fragmentation of cellular nuclear protein, with subsequent cell death (**69**).

68

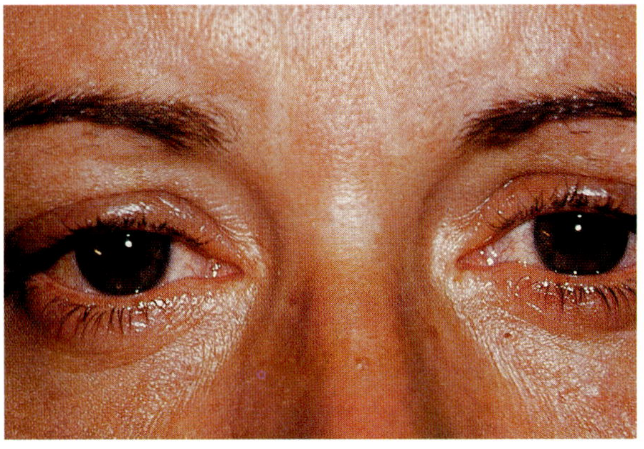

68 Skiers and alpinists must protect themselves against photophthalmia with filter lotion and sunglasses. *Courtesy of J. Figueras.*

69

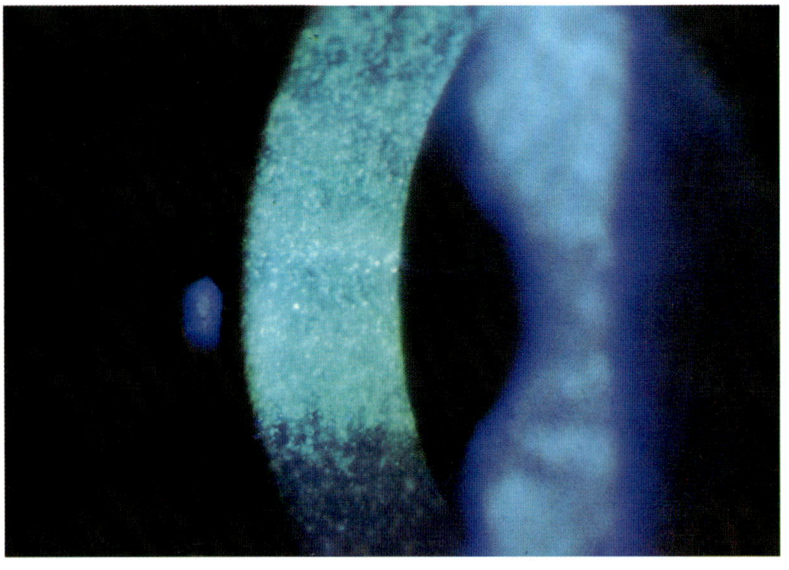

69 Excessive ultraviolet radiation results in corneal epithelial breakdown or erosion. After application of fluorescein dye, and illumination with cobalt blue light, diffuse punctate staining of the cornea can be seen. This type of corneal staining is depicted here.

Acute photophthalmia

The cornea is covered with epithelial cells. These begin to die within hours following excessive ultraviolet irradiation. This is followed by spontaneous cellular fragmentation and loss of surface adhesion. The cells are brushed off the cornea and conjunctiva by the mechanical action of blinking the eyelids, exposing the corneal nerve endings.

Contact of the blinking lids with the bare corneal nerve endings leads to symptoms ranging from a gritty, sandy sensation to excrutiating pain and sensitivity to light. The slightest movement, or exposure to light, causes involuntary spasm of the eyelids and brings on an acute paroxysm of pain. The irritated corneal nerves also cause a reflex pupillary constriction and spasm of the ciliary muscles, resulting in eye pain and headache. All of these symptoms typically occur 4–10 hours after exposure (Pitts and Tredici, 1977). The acute symptoms often last 6–8 hours, and most of the discomfort disappears within 48 hours after onset, although more severe exposure may lead to a persistent headache and some visual disturbance for several weeks. Victims of photophthalmia usually present with swollen, reddened, and occasionally blistered, eyelids and face, swollen conjunctiva with engorged blood vessels, profuse secretion of tears, a dull appearing corneal surface reflex, and constricted pupils (Duke-Elder *et al.*, 1972; Pitts, 1970).

The treatment of acute photophthalmia includes patching the eyes, and the applications of cold compresses to reduce discomfort. Dilating the pupil with atropine (1%) or homatropine (5%) relieves the ciliary spasm and reduces light sensitivity. An ocular antibiotic ointment should be applied prior to patching, to diminish the chance of developing a corneal infection before the corneal surface has healed. Analgesics and/or sedatives may be necessary.

Chronic photophthalmia

The occurrence of chronic photophthalmia has been observed among the inhabitants of mountainous and snow-bound regions, and in workers such as welders, who are frequently exposed to light sources that are rich in shortwave energy. Those inhabitants of mountainous or arctic regions who develop this disorder often experience visual disturbances and sensitivity to light, in addition to chronic conjunctival inflammation. Their conjunctiva may lose elasticity and luster, and become pigmented. The cornea may be infiltrated with a yellow-white deposit, with a bandlike distribution conforming to the exposed opening between the eyelids.

Prevention

Recognition of the harmful potential of intense sunlight, and the subsequent development of ultraviolet blocking lenses, has significantly reduced the occurrence of photophthalmia. Inhabitants of arctic and mountainous regions have long been aware of the problem and have devised primitive, but effective, protective methods. Eskimos use a leather thong to strap a piece of wood or horn, containing viewing slits, to the head. Sherpa guides tend to pull their hair forward over the eyes and face for protection.

Today, industrial exposure and tanning parlors are more frequent sources of photophthalmia than is the mountain environment for the experienced mountaineer. The source of the ultraviolet radiation is inconsequential; the intensity and length of exposure are the primary determinants of tissue damage.

Eyeglasses and goggles made to filter out or reflect wavelengths from 250–400 nm protect the cornea and interior structures of the eye from the damaging effects of excessive exposure (Pitts *et al.*, 1977). Ordinary glass provides some shielding from harmful radiation, but is insufficient to protect the eye in high dose environments. Many clear or tinted plastic lenses offer little ultraviolet filtration. Thus, attention to this detail is necessary when providing protective eye wear.

Sunburn

Damage to the skin can result from overexposure to ultraviolet rays of around 300nm wavelength. The reactivity to sunlight varies considerably from one person to another; only heavily pigmented Negroes are completely immune to skin damage. Many individuals, such as blonds and redheads, are especially susceptible and learn to avoid excessive exposure or to take appropriate safeguards to protect the skin. The effect of sunlight on the skin may also be modified by genetic, chemical, allergic or nutritional factors (**70**).

70

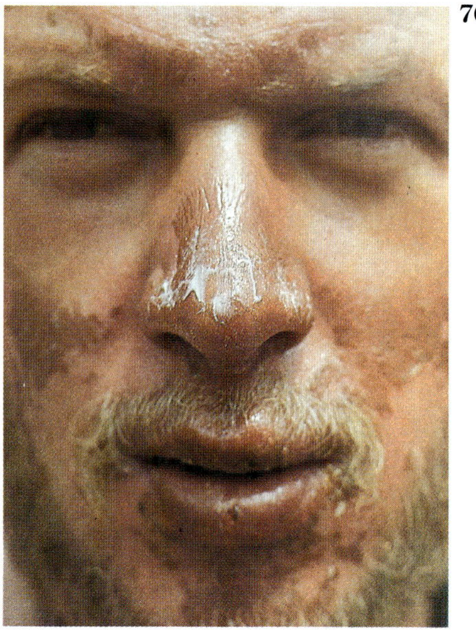

70 Acute sunburn. *Courtesy of Z. Ryn.*

Acute sunburn

Depending on the intensity of the exposure, the symptoms and signs of sunburn appear in 1–24 hours and may vary from slight erythema to severe blistering. The severity of the reaction is usually maximal after 72 hours. Mild erythema is followed by only transitory scaling, sometimes accompanied by itching. With prolonged exposure to high intensities, tenderness, pain and swelling of the exposed skin occur and extensive blistering may follow. Burning of the front of the legs is particularly painful and is usually slow to heal. Where the sunburn involves large areas of the body, general manifestations such as fever, chills, weakness and shock may occur. Such systemic responses may be indicative of heat stroke, which should be suspected, particularly if the degree of sunburn is not too severe. Following exposure to ultraviolet light, the epidermis thickens and the rate of deposition of melanin is increased, protecting the skin against further damage.

Further exposure should be avoided until the acute response has subsided. Cold compresses are helpful in relieving the acute symptoms, and creams containing corticosteroids reduce the erythema and promote healing. Ointments or lotions containing local anaesthetics should be avoided, as these are prone to lead to sensitization of the damaged skin. Secondary infection is the major complication of severe sunburn and should be treated with appropriate systemic antibiotic therapy. After severe, extensive damage to the skin, rapid improvement may be afforded by systemic corticosteroid therapy.

Chronic effects

In many individuals, long-term exposure to strong sunlight causes atrophy of the skin and produces a marked aging effect. More seriously, solar radiation can induce precancerous lesions (keratoses) of excessively exposed areas of skin; ultraviolet radiation is considered the dominant risk factor for skin cancer (squamous and basal cell carcinoma, malignant melanoma), based on the evidence of the lesions' tendency to arise on surfaces exposed to the sun, the high frequency among outdoor workers, the inverse correlation between the incidence of skin cancer and the domicile latitude, and data from experimental animal studies (Scotto *et al.*, 1982). The current fetish for sun worship in many countries is leading to an alarming increase in the incidence of these skin cancers. Prophylactic surgical or laser removal of all such lesions is preferable, but ionizing radiation and chemotherapy may also be employed.

Prevention

Severe sunburn is essentially preventable with simple precautions. Exposure to sunlight should initially be graded, and limited to 30 minutes at midday, even in dark-skinned individuals, until the skin has become pigmented at the beginning of the season.

There is a bewildering variety of sunscreening ointments, creams and lotions available. These contain molecules that absorb ultraviolet light in the part of the spectrum that can damage the skin (280–320 nm). The benzophenones can absorb up to 400 nm. Barrier sunscreens contain opaque pigments, such as titanium dioxide or zinc oxide, which scatter, reflect or block sunlight. Other compounds, which absorb light in the harmful spectrum, include *p*-aminobenzoic acid esters, benzophenones, cinnamates and anthranilates. These preparations are available without a prescription, and their labels should carry a sun protective factor (SPF) rating. This is the ratio of the time required to produce erythema through the sunscreen, to that required to produce comparable erythema without it. Current SPF ratings vary from a minimum of two up to 15 or greater.

Certain drugs can make the skin sensitive to the effects of solar radiation. These include sulfonamides, phenothiazines, dimethylchlortetracycline, thiazide diuretics, griseofulvin and quinidine. Patients taking these medications need to exercise particular care during their initial periods of exposure.

In addition to the acute and chronic effects of ultraviolet radiation, to which all white-skinned individuals are susceptible, unusual reactions can occur after brief exposure in patients suffering from lupus erythematosus – either the chronic discoid or the acute systemic type. Any patient showing pronounced photosensitivity should be investigated for this disease. Certain serious rare conditions, such as albinism, xeroderma pigmentosum and porphyria, are associated with unusually severe reactions to sunlight, and patients with these conditions are particularly prone to develop skin cancer. In patients with such abnormal sensitivity to sunlight, prolonged administration of antimalarial drugs such as chloroquine (250 mg/day po), may reduce or suppress the hypersensitivity, provided that they take care not to exceed their tolerance to sunlight. Also, those susceptible to herpes simplex ('cold sores') may experience severe exacerbations of the lesions following exposure to strong sunlight.

Eye protection at high altitude

Dominique Petetin

A great variety of sunglasses are currently available, but few are able to provide effective protection against the high level of solar radiation in the mountains and in the hot climates of seaside resorts, deserts, etc.. The eyes may be harmed by direct or reflected ultraviolet and infrared rays of the sun.

Lenses

The absorbancy of lenses

The lenses of sunglasses are classified according to their ability to absorb visible light: 'sun lenses' transmit less than 80% of visible light (high absorbancy), whereas 'filter lenses' transmit more than 80% (low absorbancy). In most cases, the lenses of sunglasses also have a higher absorbancy in the ultraviolet part of the spectrum than do 'ordinary' lenses, and some absorb infrared rays as well.

Sunglasses protect the eyes from the dazzle caused by visible light, which reduces one's sharpness of vision; erythropsia (a visual defect, in which all objects appear to have a red tinge), caused by visible light and, in particular, by green light; conjunctivitis; snow blindness, etc., caused by ultraviolet rays below 313 nm, and infrared rays above 760 nm.

When choosing sunglasses, the quality of the lenses is therefore very important (**71**). Two materials are currently used: mineral glass and organic glass (resin). Whichever is chosen, the lenses must allow clear vision, without distortion, to avoid eye fatigue. The main advantage of organic lenses is that they are unbreakable and lightweight. However, they do not filter the sun's harmful rays (particularly ultraviolet) effectively. Mineral lenses are heavier and filter ultraviolet rays better, but are breakable. They must therefore be toughened for the safety of the wearer.

The ideal lenses to wear in the mountains are treated on both surfaces. The inside of the lenses have an anti-reflective coating, to avoid the reflection of the light entering from the sides onto the cornea, and their outside surfaces are mirror finished on the lower and upper parts (to reduce both direct transmission and reflection). The upper and lower parts of the lenses are darker than the central parts, which must be light enough to allow the wearer to see clearly.

71

71 Great care must be taken when choosing sunglasses, especially for children.

Frames

When choosing frames, the wearer should select the most suitable shape for his face. The frames should not rub too heavily on the cheeks, as this may cause the skin to chap. Side and nasal masks are essential, to provide additional protection against reflected light. The temple pieces of the glasses should be adjustable and snug, to prevent the frames from slipping during movement. The frames should be lightweight, but able to resist shock and temperature fluctuations. A safety cord must be fixed to the temple pieces of the glasses, and worn around the neck. A spare pair of glasses is essential.

Cold

Frostbite and mountaineering

Bernard Marsigny and Jacques Foray

Frostbite is a localized injury, caused by the direct action of coldness during a relatively long exposure to a temperature below 0°C. It results from the vasomotor response to coldness, which induces ischemia and consequently aggravates the effects of the cold.

Etiology

Four principal factors, together with **coldness**, are responsible for the appearance of frostbite:

- **Humidity** multiplies the effect of coldness by a factor of 14.
- **Wind** multiplies it by a factor of 10.
- **Altitude polyglobulia and dehydration** contribute to the slowing down of the capillary blood flow.
- **Exhaustion.**

Clinical symptoms

Frostbite mainly effects young people, mountaineers (59%) and skiers (28%). Of the 1,085 people treated for frostbite in the Surgical Department of the Chamonix-Mont-Blanc Hospital (over a 20-year period), it has been noted that 57% had frostbite of the feet, 46% of the hands, and 17% of the face (**72**).

There are three consecutive stages in the development of frostbite. The first is a painful feeling of coldness, with the classic tingling and aching of numbed fingertips. The second stage is insidious, as the affected parts of the body become anesthetized, allowing the injuries to settle. The third stage occurs as the patient warms up, with the reappearance of painful pheno-

72

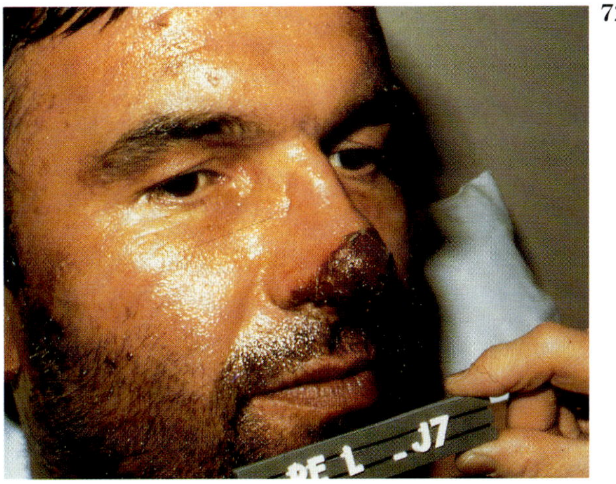

72 Third degree frostbite of the nose; prevention plays a vital role.

mena, followed rapidly by a more or less significant edema with clear or serohematic blebs. Necrosis does not appear until much later.

As with burns, frostbite is classified in three degrees, according to the various clinical tests and the prognosis (**Tables 2a–c; 73–80**).

Table 2a Classification.

Diagnostic classification	*Evolution*	
Superficial frostbite	First degree	41%
	Second superficial degree	33%
		74%
Deep frostbite	Second deep degree	18%
	Third degree	8%
		26%

Table 2b Superficial frostbite.

Symptoms	*Evolution*
First degree	
Paleness, then erythema during warming up	Complete recovery
Cyanosis (decreases rapidly)	
Edema + or –	
Blunted sensitivity	
Second superficial degree	
Persistent erythema after warming up	Slow but complete recovery
Persistent cyanosis	
Edema	Possible sequellae:
Clear blebs	Hypersensitivity to cold
Decreased sensitivity or anesthesia	Neurovascular microlesions in the fingertips

Table 2c Deep frostbite.

Symptoms	*Evolution*
Second deep degree	
Paleness, then significant cyanosis	Derm-limited necrosis
Edema ++	Slow recovery of sensitivity
Serohematic blebs	Neurovascular microlesions in the fingertips
Anesthesia	Recovery after four or five weeks with sequellae
Third degree	
Persistent paleness and cyanosis, then the appearance of deep necrosis	Deep necrosis (muscle and bone)
Edema +++	Amputation
Anesthesia	Significant sequellae

73 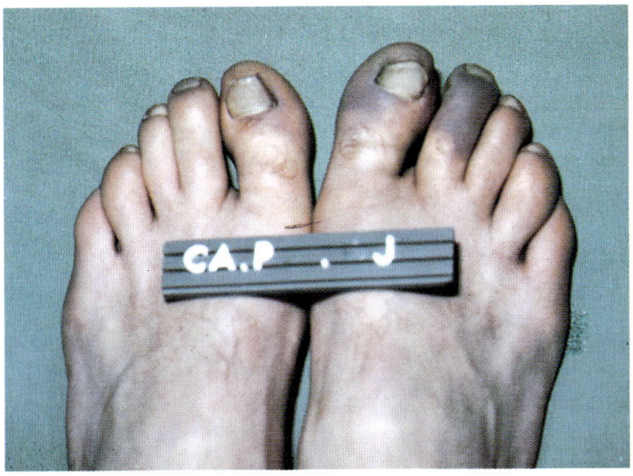

74 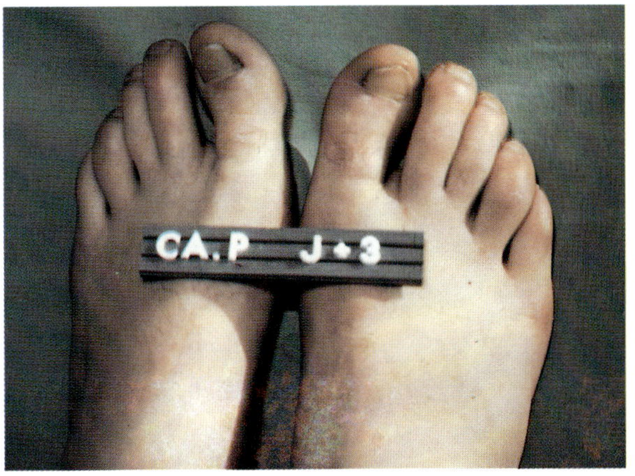

73 First degree frostbite on the day of arrival.

74 First degree frostbite in the same patient, healed on the third day.

75 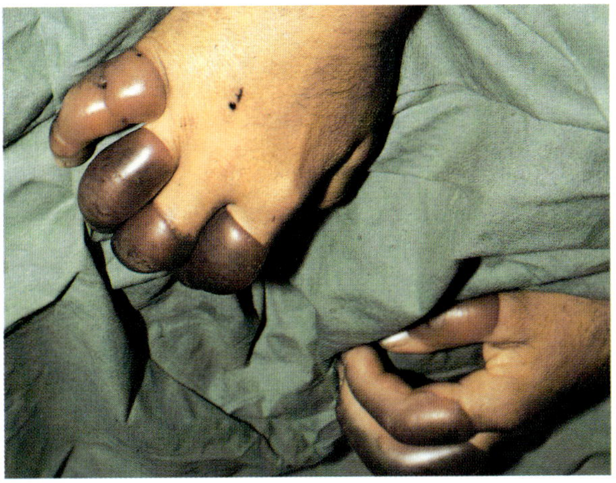

76 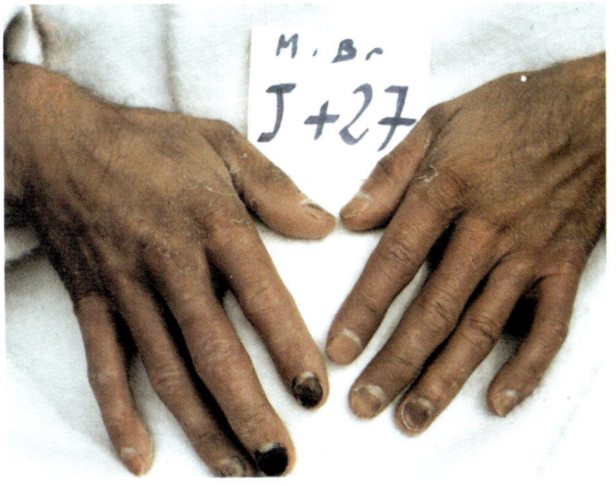

75 Second superficial degree frostbite, second day.

76 Second superficial degree frostbite in the same patient, day 27.

77 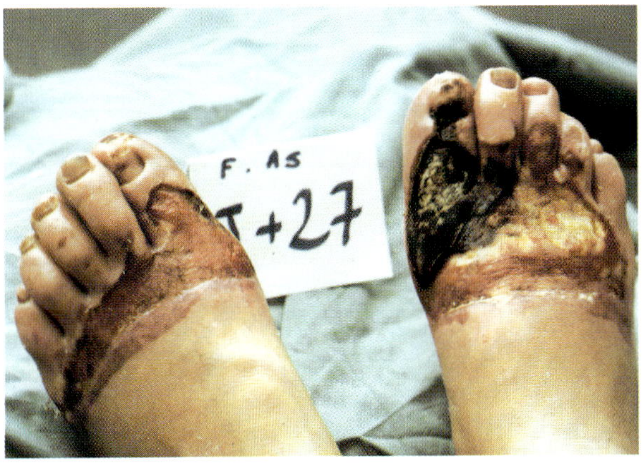

78 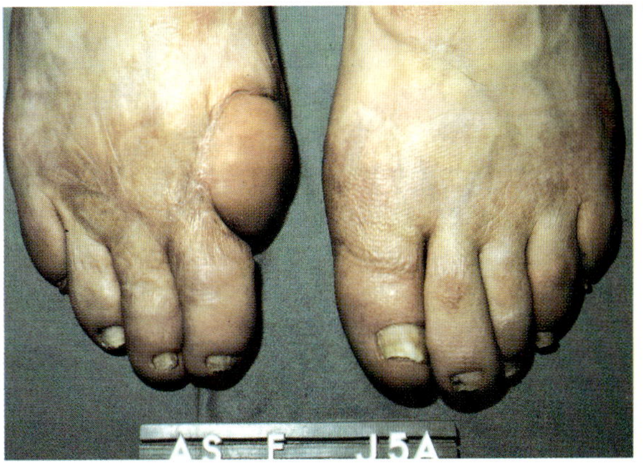

77 Frostbitten feet: the big toe of the right foot has third degree frostbite; elsewhere, second deep degree frostbite is present.

78 The same patient five years later.

79

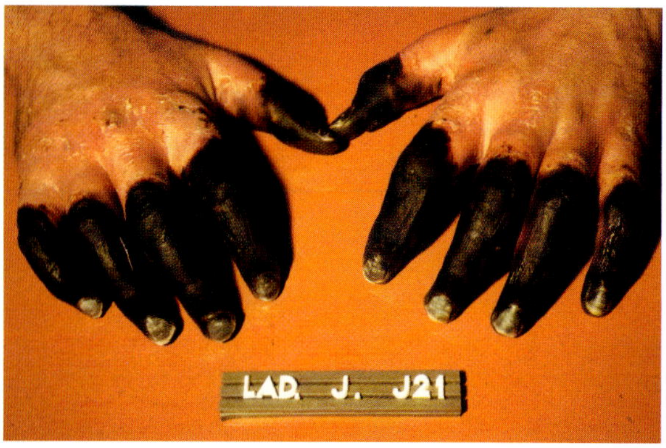

80

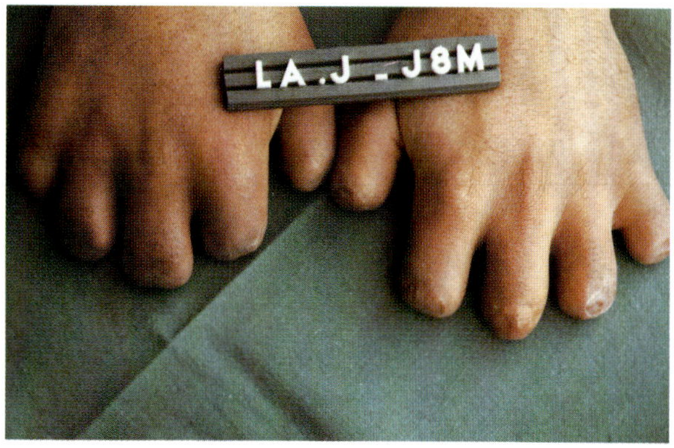

79 Third degree frostbitten fingers.

80 The condition of the same patient after eight months.

Prognosis

Early prognosis of frostbite is difficult to establish clinically. In fact, one has to wait four or five days to know whether the frostbite is superficial or deep. Where the frostbite is deep, one has to wait approximately 45 days for the appearance of the demarcation limit, to establish an optimal level of amputation. Fortunately, some complementary forms of examination can help to evaluate the prognosis.

Delayed **bone scanning** with technetium-99m is now the main diagnostic procedure at the Chamonix Hospital. Performed within two days of admission to hospital, this examination allows an early prognosis to be established, and clearly indicates which patients are likely to require amputation. Bone imaging has gradually replaced other forms of examination, such as thermography, cutaneous thermometry, arteriography, capillaroscopy and Laser-Doppler.

Nuclear magnetic resonance (NMR) spectrometry has been introduced at Chamonix Hospital, using phosphorus-31 to gain information on cellular metabolism. This method allows the early identification of necrosis and the measurement of the effect of a given therapy. This latest procedure, reliable for legs and forearms, unfortunately cannot yet be used for lightly muscled areas such as fingers and toes.

Treatment

The purpose of the treatment is to prevent secondary effects, to avoid amputation and to attain recovery as quickly as possible. The treatment must therefore be immediate, intensive, and above all, as conservative as possible.

The first step is to stop the exposure to cold. **Warming** should be rapid, but should only be started when one is sure that the injuries will not be re-exposed and allowed to refreeze. The edema that appears with the warming process renders the frozen area more vulnerable, and can prevent the mountaineer from putting on his boots again.

Vasodilators may be used in the treatment.

In addition to the classical naftidrofuryl and buflomedil chlorhydrate, the authors currently use two more recently explored drugs. One of these, nicardipine, is a calcium antagonist. The other, ketanserine, is antagonistic to the S2 serotonine receptors, and has the two complementary effects of vasodilatation and antiaggregation of the blood platelets. The authors have used it in about sixty cases, with very good results.

Blockade of the sympathetic nervous system may also be carried out, by means of a continuous brachial plexus block, or a continuous lumbar epidural block; both are easily applicable and give good pain control.

Normovolemic blood dilution increases the blood flow of the microcirculation by inducing an improvement in the rheological condition. The oxygen supply is maximal for a viscosity of 30–35%. This technique seems to be a useful complement to the vasodilators.

Anticoagulants and antiaggregants may also form part of the treatment. Dextran is used in doses of 500 cc per day. Heparin is helpful in preventing microthrombosis, and for its anti-inflammatory action.

The main complication in frostbite is a secondary infection of the wounds, and precautions must be taken **against infection**. The frozen areas should be soaked in mild antiseptic solution twice a day, and left uncovered. They should be kept in a slightly raised position on a sterile sheet. Antibiotic therapy is not systematic.

Surgery plays only a late and limited role.

Chamonix Hospital operational procedure

A vasodilator is first slowly injected intravenously. The frostbitten extremities are then soaked for 30 minutes in 37–38°C water, mixed with an antiseptic solution. There are two possible results. The wounds may heal almost completely, showing that the frostbite was superficial. In this case, the patient can be dismissed from the hospital, but should continue to have medical check-ups, and vasodilative and antiseptic treatment. Alternatively, the wounds may not heal, in which case the patient must be admitted to hospital for general treatment. The frostbitten area must be soaked twice a day, raised and left uncovered on a sterile sheet. Perfusion of dextran (500 cc per day) should be carried out for four days, and naftidrofuryl or ketanserine given intravenously. Later, the ketanserine is taken orally (3 × 20 mg per day), rather than intravenously.

Results

In the authors' experience, only 8% of cases involve third degree frostbite, and lead to amputation, which is seen as a poor result. The good results are considered to be those that do not involve amputation, but do lead to some kind of disability due to sequellae, for example, thin and cracked skin, or deformation of the small joints. Hypersensitivity to coldness is noticed in 80% of patients, but diminishes with time. Excellent results after a first degree, or a second superficial degree, of frostbite are experienced by 75% of patients. If one is ready to admit that second deep degree frostbite can evolve favorably, it could be said that early well-treated frostbite leads to good results in 80–85% of patients.

Frostbite

William J. Mills, Jr

Frostbite is true tissue freezing, and occurs when there is sufficient heat loss in the local area to allow ice crystals to form in the extracellular spaces, and extract cellular water. Freezing injury, as found in Alaska, can occur by the following mechanisms:

- True frostbite: superficial or deep.
- Mixed injury: immersion injury (wet, cold injury), followed by freezing; usually disastrous and often quite painful.
- Freezing, thawing by any means, with refreezing; again generally disastrous in result, with total tissue destruction and early mummification of distal tissues occurring within five days.
- Hypoxia, high-altitude environment injury, often with dehydration of tissues, due to general body dehydration and hypovolemia with extremity freezing; prognosis poor, especially if associated with compartment pressure syndrome (**81** & **82**).
- Extremity compartment compression, from any cause, followed by freezing; very poor results if compartment pressures are not relieved by fasciotomy.
- Extremity fracture or dislocation and superimposed freezing; the results are poor if the fracture or the dislocation is left unreduced; best results appear to follow rapid rewarming techniques.
- Hypothermia, associated with superimposed freezing injury of extremities. Paramount importance is given to the restoration of heat in the victim, under total physiological control and monitoring. Best results for freezing injury appear to be associated with tub rewarming of hypothermia, and simultaneous thawing in warm water of the frozen extremity. The danger here is the sudden release of metabolites and the release of excess amounts of potassium from muscle degradation and injury, which may cause cardioplegia. The immediate balance of electrolytes and the restoration of normal pH levels is imperative. The excellent method of rewarming with peritoneal dialysis may require almost simultaneous warming of the frozen extremity by other means (Mills, 1983; Mills *et al.*, 1960 & 1961).

81
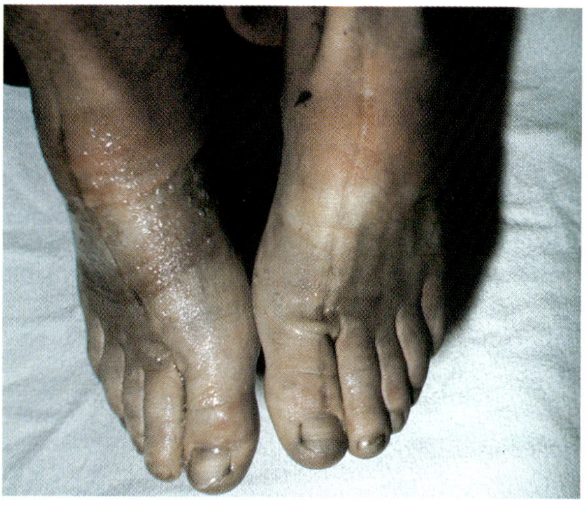

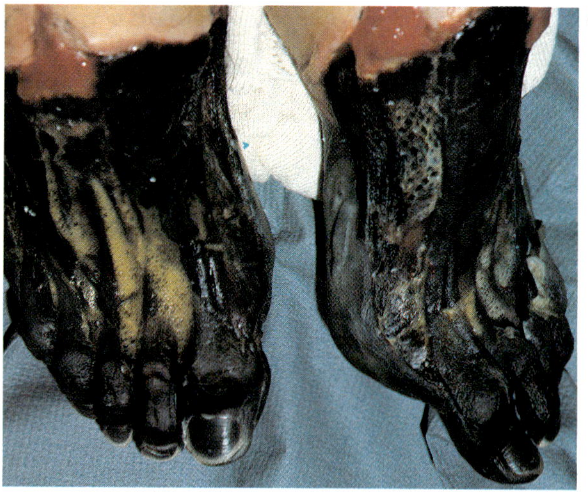
82

81 & 82 Hypoxia, a high-altitude environment injury with tissue dehydration, severe freezing of the tissues with the imprint of a neoprene sock with zipper. This is a freeze–thaw–refreeze injury of great severity which, despite fasciotomy and other care, resulted in demarcation across the proximal tarsi. The injury was sustained between 5,200 and 5,800 m, on Mt. McKinley, where the patient was without food or water for a least three days. Irrevocable changes were apparent from the moment of thawing.

Physiology

Present knowledge would indicate that the pathophysiological changes occur in two stages. First are changes occurring in and induced by the freezing state, namely: structural damage by ice crystal growth; protein denaturation; pH changes (intercellular and extracellular); dehydration within cells; loss of protein-bound water; rupture of cell membranes; abnormal cell wall permeability; destruction of essential enzymes; ultrastructural damage to capillaries; consistent mitochondrial damage in muscle cells.

During the thaw and post-thaw stage, the changes may include: circulatory stasis; corpuscular aggregation; venule obstruction; piling of red cells back to capillary bed; development of hyaline plugs in the vascular tree; marked tissue edema; anoxia-ischemia of tissues; increased compartment space pressure; capillary and peripheral vessel collapse with, eventually, thrombosis, ischemia, regional necrosis and tissue death, if the process is not reversed (**83** & **84**) (Meryman, 1957 & 1966; Karow and Watts, 1965; Quintanella *et al.*, 1947; Lange *et al.*, 1945).

A more recent theory has been developed, that breakdown products of arachidonic acid are mediators of progressive dermal ischemia. The arachidonic acid cascade, primarily thromboxane and prostaglandins, may permit increased vascular clotting (MacCauley *et al.*, 1983).

83

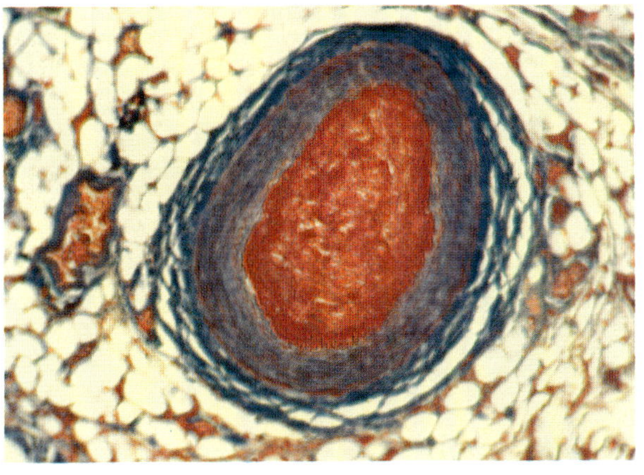

83 A magnified view of the arterial wall, with total unorganized thrombus and extensive necrosis.

84

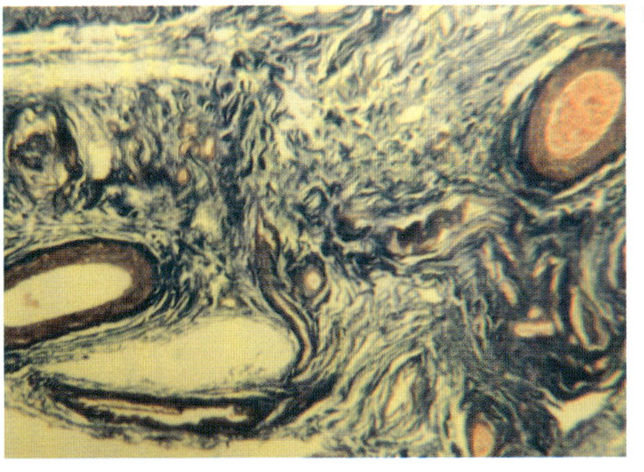

84 This pattern is typical of a freeze–thaw–refreeze injury. Demonstrated are occluded small vessels, and devitalized structures, including nerve necrosis. Loss of intimal lining, subintimal necrosis and fracture of normal elastic fibers are present.

Contributing factors

New trends may establish changes in injury (**85**). For many years in Alaska, freezing of ears, often with tissue loss, was not uncommon, especially in male children. The injury occurred in skiers, skaters and snowmobilers. With the advent of longer, neck-length hairstyles, frostbite of the ear was seldom seen for almost a decade. However, with the return of the short hairstyle, the frozen ear pattern is with us again (Mills, 1983; Mills *et al.*, 1960 & 1961).

In recent years, snow boots with felt liners have been popular. When the foot is immersed in water, or the felt wetted by some means, such as melted snow, the felt liner may shrink, contract and freeze. Extremity freezing may follow, complicated first by vascular occlusion, as the contracted felt liner acts as a tourniquet. Similarly, neoprene or rubber scuba boots used by the mountaineer may cause occlusion of circulation at high altitudes. This is because of pressure changes at lowered atmospheric pressure. Freezing following this pre-existing vascular occlusion greatly complicates the injury and the final result (Mills *et al.*, 1977; Sutton *et al.* 1987).

Similarly, running shoes or tennis shoes allow freezing injuries to develop. Even in Alaska, in coastal or interior areas, regardless of the low temperature, the wind and snow depth, Alaskan students and many adults risk freezing and non-freezing injury through wearing inadequate footwear. Their injury toll is almost matched by that of the cross-country skier, competing in low temperatures or backpacking in wilderness areas wearing ultralight low-cut shoes with toe clips.

Alcohol and drug abuse contribute to hypothermia and freezing injury by impairing mental and physical function. Recently, it has become apparent that the nasal 'snorting' of cocaine, by causing constriction of mucous membranes and the nasal arterial supply, has allowed serious frostbite to occur in the nose – an area usually so well supplied with blood that deep injury was considered rare.

85

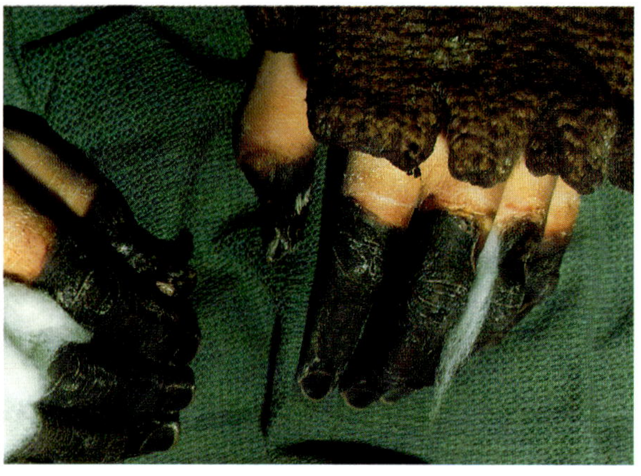

85 Fingerless mittens that leave the thumb and four digits uncovered are often used for work, particularly by alpine photographers, in cold weather or at altitude. Unfortunately, in severe cold, at altitude and in high winds, freezing injury occurs rapidly, and recurrent use of this type of glove allows a freeze–thaw–refreeze injury to develop. The demarcation and gangrenous changes pictured here were suffered by a climber injured near the summit of Mt. McKinley, with mummification occurring by the fifth day; this is almost pathognomonic of multiple insult and freeze–thaw–refreeze injury.

Rewarming procedure

In order of prognosis, from best to worst, methods of thawing are (Mills, 1983; Mills *et al.*, 1960 & 1961):

- Rapid rewarming in warm water (37–41°C) (**86–88**).
- Gradual, spontaneous thawing at room temperature (the problem here being the difference in room temperature between an average, heated home and a cool cabin in the wilderness).
- Delayed thawing, using ice and snow techniques.
- Thawing by excessive heat (50°C or higher) (**89**).

At present, rapid rewarming in warm water is favored, as this method seems to demonstrate the greatest tissue preservation and the most adequate early function, especially in deep injury. Results of gradual, spontaneous thawing vary in deep injury, but seem satisfactory in the superficial injury patients. Ice and snow thawing gives variable results, most often poor, with marked loss of tissue. The use of excessive heat as a thawing method has resulted in disaster in most cases, especially dry heat at temperatures of 66–82°C, for example, the use of diesel exhaust, wood fire or stove heat.

Much controversy exists regarding thawing methods, both in North America and in Europe. Certain conclusions may be drawn from 1,120 cases of cold injury, treated over the last 20 years (frostbite: 903; hypothermia: 133; immersion injury: 84 – of this group, 51 cases were hypothermia and frostbite combined, while 20 cases were immersion injury, complicated by freezing injury) (Mills, 1983; Mills *et al.*, 1960 & 1961; Ariev, 1955; Killian, 1952; Foray *et al.*, 1977; Campbell, 1964).

86

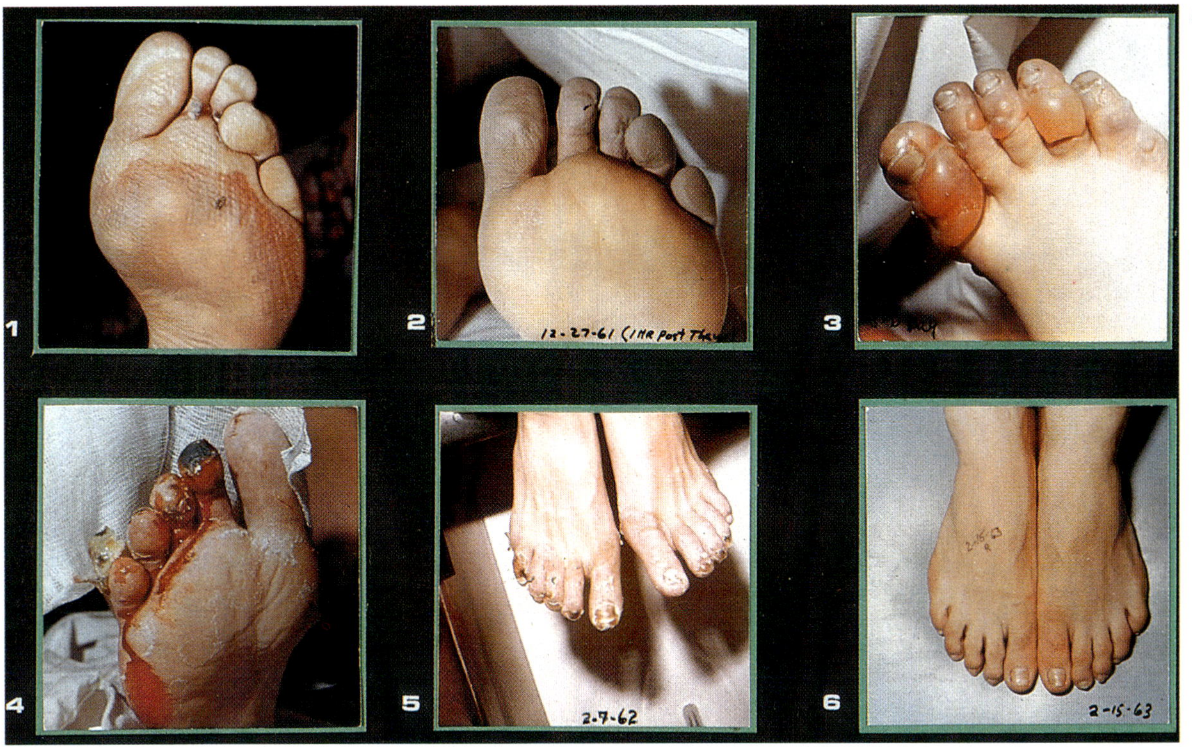

86 Rapid rewarming in a warm water bath.

Frame 1 A young boy, aged 14, was exposed to temperatures of –20°C with very little wind. He was wearing leather boots that were too tight. His exposure time was six hours. He arrived home with his overboots filled with snow, and frozen to the leather boots underneath. He thawed his feet in a warm water bath, approximately 43°C, until the tips of his toes flushed. The slide demonstrates a cold, rigid forefoot. Tissue compression and stocking marks are clearly visible on the skin. The toe joints were immobile and insensitive.

Frame 2 Thawing was followed by the development of a burgundy discoloration. Much less cyanosis is present if the thawing temperature is kept in the range of 32–38°C, until the tips of the toes or of the fingers flush.

Frame 3 Over the next 48 hours, large, clear blebs developed, ultimately extending to the tips of the toes, except for the second toe.

Frame 4 Revascularization of the tips of the toes occurred with epidermal shedding, except the tip of the second toe which was closed by epithelialization.

Frame 5 By the sixth week, desquamation is complete, with the loss of the nail of the large toe (left).

Frame 6 One year post-injury, the gross anatomy is intact, but the changes caused by deep injury are obvious. These include volar fat pad loss, subcutaneous fat pad loss, early interphalangeal joint contracture, hypesthesia of the tips of the toes and hyperhydrosis. Sensation was improving progressively.

87

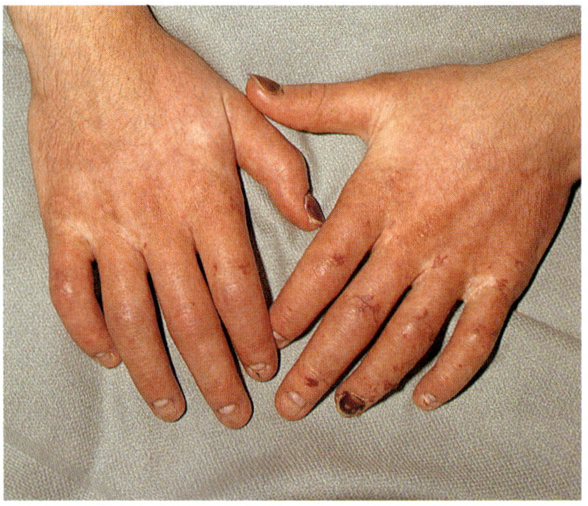

88

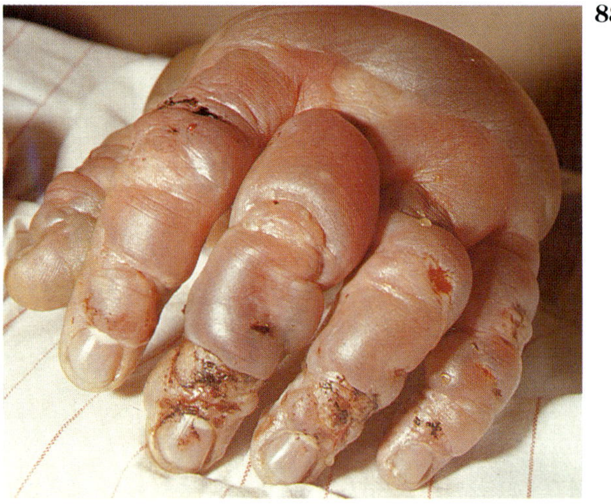

87 & 88 Bleb formation 24 hours post-thaw. The patient has undergone rapid rewarming in warm water for approximately 30 minutes, until the fingertips flushed. It should be noted that the blebs are large, pink, and extend right to the fingertips – an excellent prognostic sign. Until the blebs developed, following the rapid rewarming, this patient still had feeling in his fingertips, which disappeared as the blebs separated the superficial from the deep structures.

Many thawing methods are utilized by the victims, or by helpful rescuers. Not all are appropriate. Generally, the victims present to the rescue areas or emergency rooms, or are treated with one of the methods of thawing discussed below (again in decreasing order of effectiveness).

(i) **Rapid rewarming in warm water (32–41°C),** in a tub or a whirlpool bath, or by means of a crane lift platform in a Hubbard tank (**90**). Rapid rewarming by external means appears to produce the best results, but does not always give protection from tissue loss, especially where the injury is deep or of long duration.

(ii) **Gradual, spontaneous thawing,** at room temperature (a variable accumulation of temperatures from 7–32°C), due to cabin heat, or thawing occurring in travel by foot, by vehicle or during rescue; thawing due to the warmth of a sleeping bag, often in the wilderness or at altitude. Spontaneous thawing gives variable results, which are often determined by the depth of injury, the duration of freezing, and the patient's activity during survival, rescue and thawing.

(iii) **Delayed thawing,** using ice and snow techniques, cold water and friction massage. 'Warming' by cool methods (usually at temperatures near freezing) in warm areas, often gives poor results.

(iv) **Thawing by excessive heat,** such as campfire heat, oven heat or engine exhaust (temperatures greater than 50°C). This method generally results in heat injury (burning) to a part already injured by cold. The results are disastrous, often resulting in major amputation.

89

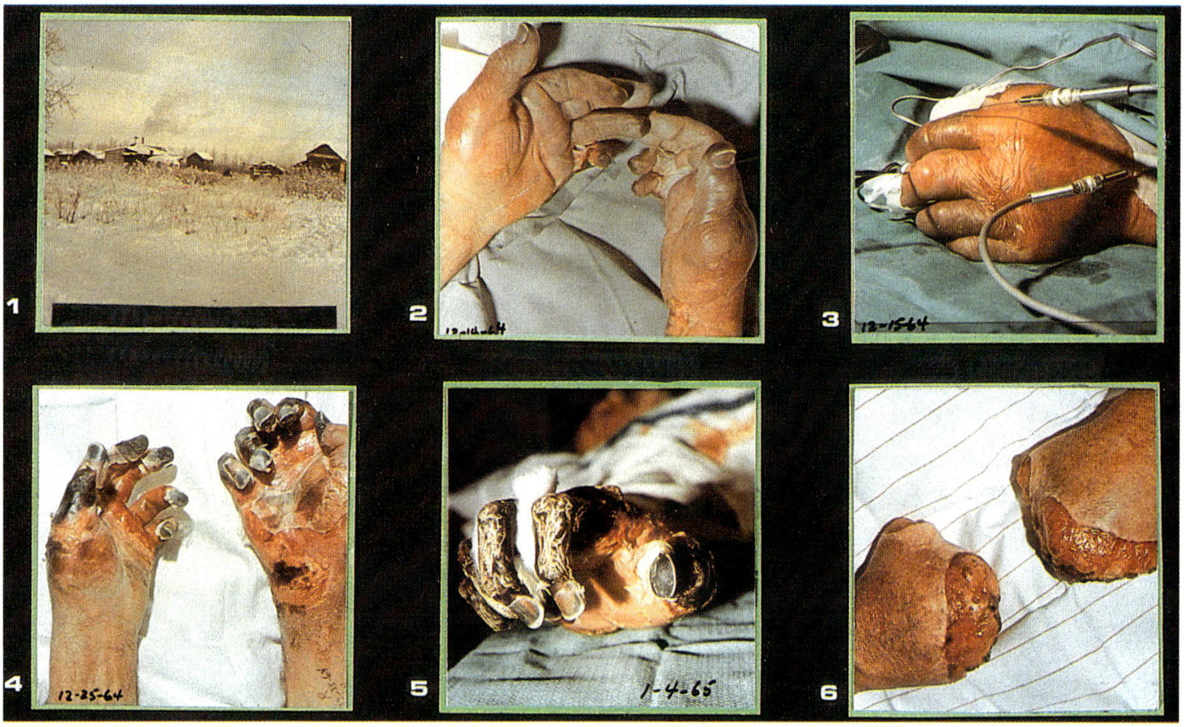

89 Thawing by excessive heat – tea kettle water.
Frame 1 A 70-year-old fisherman was exposed to a temperature of –27°C for five hours. He eventually crawled to a village in Alaska, where his hands and feet were found to be frozen solid. The hands were hard, cold, frozen and white, curled into full fists, club-like. Thawing was induced firstly by means of ice and snow and massage. and then by using simmering tea kettle water, thereby 'burning' the frozen parts at temperatures greater than 70°C.
Frame 2 The burn injury, superimposed on the previous freezing injury.
Frame 3 A thermistor evaluation of the hand temperature on the third day demonstrates a drop in tissue temperature (deep) to 21°C. This evaluation is now performed using radioisotopes.
Frame 4 By the twelfth day, total necrosis and digital mummification are apparent, with severe injury to the metacarpophalangeal junction.
Frame 5 Total destruction and mummification of tissue is present at three weeks. This early mummification, appearing by day five to seven, is usually found in a freeze–thaw–refreeze injury, or after thawing by excessive heat, or following immersion injury with freezing. Superficial infection is present at the demarcation line, but is held in abeyance by whirlpool washing which dilutes the bacterial content. This is important, as the loss of normal tissue and of the vascular transport system prohibits systemic antibiotics from reaching the area of demarcation.
Frame 6 At six weeks, spontaneous amputation is identified bilaterally, to the level of the metacarpophalangeal junction. This result demonstrates the poor chances of recovery following necrosis and gangrene, which are found when frozen tissues are warmed or 'cooked' by excessive dry or wet heat.

90 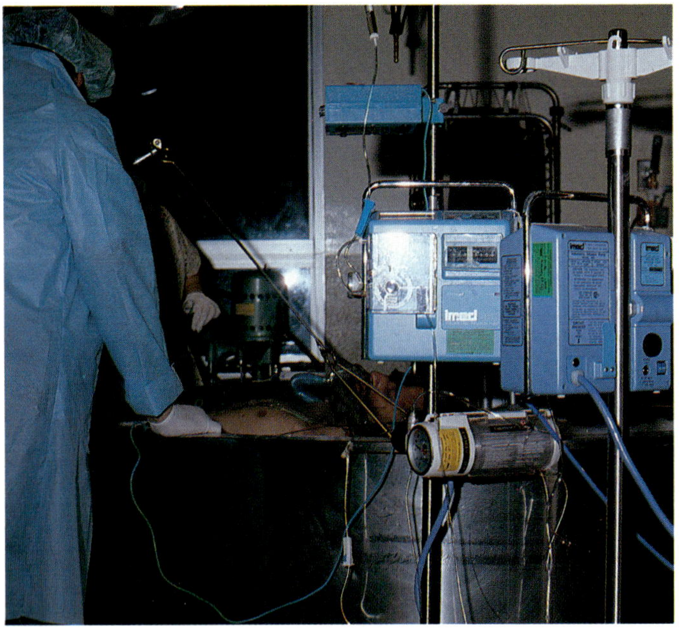

90 Peritoneal dialysis is continued as the patient is placed in a warm tub for rapid rewarming of the extremities, when his core temperature reaches 27°C. The patient's deep hypothermia and multiple episodes of freezing and thawing, resulted in marked, overwhelming sepsis which required emergency bimalleolar amputation of the feet; the hands fared better with minimal phalangeal loss.

An early diagnostic clue as to the exact freezing event is the condition of the digital tips when the patient is first seen. If spatulate mummification occurs within the first three to five days after thawing, the diagnosis is usually a freeze–thaw–refreeze injury (**91**), or thawing by excessive heat. Freezing of tissues, with thawing by means other than freeze–thaw–refreeze or excessive heat, usually demonstrates tissue demarcation, necrosis or mummification in 10–21 days, not five days.

It has been stated that rapid rewarming by internal means (warm intravenous fluids or arterial line fluids), at temperatures of 37–41°C, is more physiological, and may be a method of choice in dissolving ice crystals and restoring cellular hydration. Though this method appears more logical and is a new consideration on the horizon of care, it has, in fact, been a method of choice for over 10 years, at least in this area and elsewhere, for adding heat and restoring fluid volume with the heated solutions. However, the results are still no better than those produced by rapid rewarming methods. In addition, the development of an arterial line, especially in the areas of the ankle and wrist, may cause local arterial spasm and further decrease digital perfusion. The ideal method is not yet at hand (at least for thawing of the frozen part), but tissue loss is less now than in past decades, regardless of thawing methods. Major above-knee or below-knee amputations, or amputations at wrist or forearm level, are far fewer in number (Mills, 1983; Mills *et al.*, 1960 & 1961; Ariev, 1955; Killian, 1952; Foray *et al.*, 1977).

91

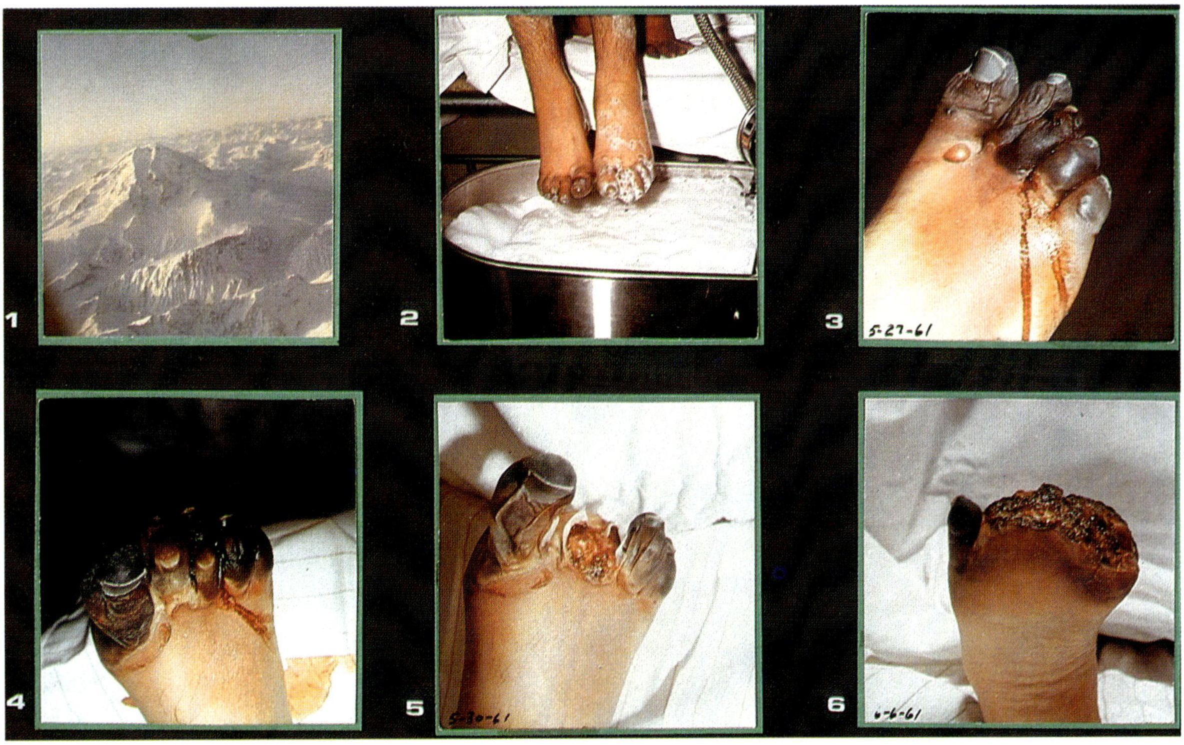

91 The mountaineer's dilemma: freeze–thaw–refreeze.

Frame 1 A Mt. McKinley climber, aged 31, together with his companion, sustained freezing of both the hands and the feet, at an altitude of 5,182 m and at a temperature of between –26 and –34°C, with 15–20 knot winds. Each climber's frozen parts were rewarmed by body heat (axilla and groin). Two days later, the climber's swollen and edematous feet were placed back in 'tight' boots. As they were very close to the summit, they continued their climb. Soon after reaching the peak of Mt. McKinley, where the temperature and the wind were roughly the same as on the day of freezing, both noticed a total loss of sensation in their fingers and toes. The second thawing was carried out in a similar fashion to the first. The climbers then walked down to 4,267 m and were airlifted by helicopter from that level.

Frame 2 48 hours post-second thaw, the patients were placed in a whirlpool bath for continued warming. Here, the distal areas and toes are cyanotic and dark, and demonstrate small proximal serial sanguinous blebs, a poor prognostic sign.

Frame 3 On the fifth day post-thaw, the foot is wet and edematous, and insensitive, with early forefoot necrosis. Early demarcation between viable and nonviable tissue is apparent.

Frame 4 Liquefication necrosis and mummification are present by the third hospital day, with the obvious early infection developing.

Frame 5 By the seventh day post-thaw, rapid liquefication necrosis of the forefoot, and initial separation of tissues with dissolution of ligaments, vessels, nerves, tendons and structures about the joint are demonstrated. The osseous structures are held in place only by a skin envelope.

Frame 6 On the twelfth day post-thaw, there has been self-amputation of the toes without surgery. This can result from tissue injury following freeze–thaw–refreeze injury or immersion injury, followed by further freezing. This injury, up to the present date, has proved to be irreversible. The tissues are often lost at the level of the second freeze.

Treatment

Treatment can generally be divided into two categories: **pre-thaw** and **post-thaw**.

Pre-thaw treatment

The frozen part must be protected to avoid trauma and the risk of irreversible injury at the frozen–non-frozen interface, which may result if motion occurs at that level, fragmenting partially frozen tissue. The frozen part should then be thawed in a whirlpool bath or a tub or, if nothing else is available, with warm wet packs at 38–41°C. At Providence Hospital in Anchorage, Alaska, an electrically operated hoist (crane) has been in use since 1963, which lowers the patient's whole body into a Hubbard Tub, with whirlpool. This method has also been used for the rewarming of hypothermia victims, and to warm and thaw the combined injury of hypothermia with extreme freezing. Temperatures should not exceed 41°C. The thawing is completed when the distal tip of the thawed part flushes. Sedatives or analgesics may be utilized if the thawing process is painful and cannot be tolerated. The thawed part should not be massaged. Rapid rewarming must not be used if the part has been thawed previously.

Post-thaw treatment

When injury is severe and deep, and hospitalization is required, the extremities are kept on sterile or clean sheets, with cradles over the frostbitten part to avoid trauma and pressure. This is not necessary for the upper extremities, which may be laid out on sterile sheets placed over the chest and trunk. Treatment is open, not occlusive, without the use of wet dressings, unguents, ointments or petrolatum gauze. Whirlpool baths are utilized twice daily for 20 minutes at a time, at temperatures of 32–35°C. Surgical soaps, such as hexachlorophane, 4% chlorhexidine gluconate, and povidone iodine, are employed in the whirlpool. Occasionally, following Moyer's methods for burns, 0.5% silver nitrate solution may be lavaged over the area of frostbite. The end result is similar to that produced by the surgical soaps; epithelialization is similar, with one outstanding difference: pain is reduced and infection, even superficial, is much less obvious using the silver nitrate solution. The whirlpool clears the debris from the injury and removes superficial bacteria. The tissues are débrided without trauma by the whirlpool action, when they are physiologically prepared for the separation of viable tissue from the overlying eschar.

Recently, 1% silver sulfadiazine solution has been utilized on open wounds secondary to freezing injury, when severe drying and premature blister rupture have occurred, followed by superficial infection. Its use occasionally prevents eschar separation, apparently by inhibiting proteolytic enzyme production.

Generally, blisters are left intact because the contents are usually sterile, as are the underlying tissues. The blisters are débrided or trimmed only if they are infected and contain purulent material. Escharotomy should be performed on the dorsum or lateral aspect of the digits when the eschar is dry and has firmed sufficiently to have a cast effect on the digits, limiting their joint motion. Digits will be débrided further in the whirlpool, without prematurely exposing underlying granulation tissues. Débridement or amputation should be delayed until sufficient time (often 15–45 days) has elapsed to demonstrate mummification and tissue death, with no danger of further retraction of tissues.

Overwhelming infection, often found in freeze–thaw–refreeze injuries, or in extremity trauma complicated by freezing, may result in overwhelming sepsis, requiring immediate amputation to avoid toxic shock.

If the extremity has remained in a frozen state for some considerable time, even rapid thawing and general supportive care may not be effective in restoring the circulation, and a condition similar to anterior tibial compartment syndrome may be clinically demonstrated. This problem may require fasciotomy. The condition can be determined either clinically, or by measuring compartment pressures, by the use of arteriography or the injection of isotopes such as technetium-99m, to demonstrate the state of cellular perfusion (Mills, 1983).

Isotope studies have been performed as a diagnostic aid of cellular perfusion for over 30 years (**92**) (Mills *et al.*, 1960 & 1961). Doppler ultrasound has also been used as a vascular study tool. Interestingly, Thermal Unit patients at Providence Hospital, with evidence of good Doppler pulses in the distal extremities (distal digital vessels), have had conflicting isotope evidence showing failure of extremity perfusion in the same

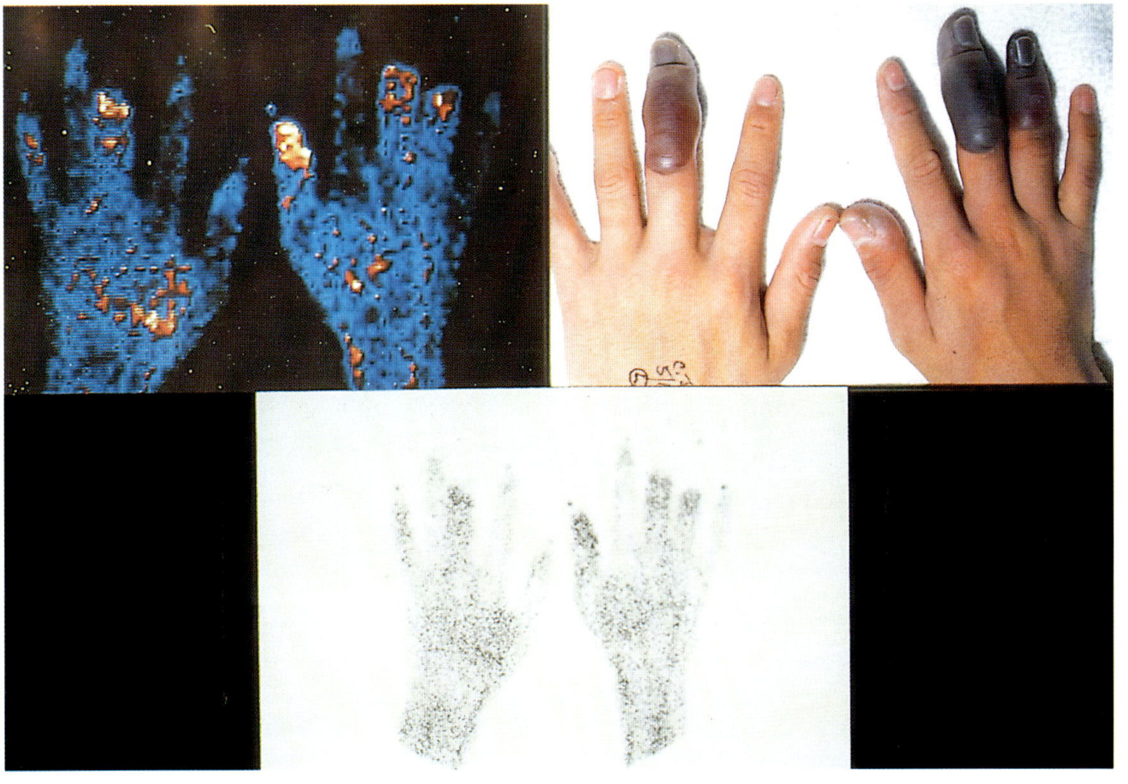

92

92 Technetium-99 m studies, combined with a thermographic pattern, are useful in the prognosis of cold injury. Demonstrated here are the isotope and thermograph studies, as compared with the cold-injured hand at that time; information is given not only on the prognosis, but also on the evaluation of drug therapy, and fasciotomy or other care. The accuracy of the isotope studies may be well over 95%. Isotope studies of cold-injured extremities are particularly favored because they are non-invasive, causing no trauma to the already injured part.

area. In all cases but one, the isotope study was the accurate one. Evidently, large digital vessels may remain patent for a short while, even when the deep capillary system is blocked. Failure of sophisticated tools has also been demonstrated in the use of devices for measuring compartment pressure. If one's clinical judgement and experience advise that an immediate fasciotomy is required, whereas the pressure transducer measuring device indicates that the pressure is high but not lethal, or indicates a marginal reading, it is often better to trust one's own judgement. A later measurement may indicate sudden pressure increase. A delay in performing the fasciotomy may be disastrous. This diagnostic problem may be avoided by the use of continuous pressure monitoring. However, there are many pitfalls. The monitoring device is still only a machine, and one's own studied, concerned opinion that the fasciotomy should be performed may preserve the limb.

The grafting of split thickness skin to large granulating areas, or areas where skin cover is considered proper, may be carried out from the third to the fourteenth day. The results of skin graft are best following thawing by rapid rewarming. The pedicle grafting of full thickness skin is a late procedure.

The use of a mesh skin graft at the time of fasciotomy or soon after, reduces the morbidity and lowers the incidence of scarring and infection.

Antibiotics are not necessary, except where infection is deep. Common bacterial organisms found in the injured tissues include *Staphylococci, Streptococci, Pseudomonas* species, and often an abundance of gram-negative species. *Clostridia* species are occasionally found. Antibiotics are not always used with the whirlpool care, but routine cultures and sensitivity studies are taken, and at the first indication of non-superficial infection, not cleared by whirlpool washing, aggressive antibiotic therapy by oral, intramuscular or intravenous methods is utilized. It is again stressed that one of the purposes of the post-thaw whirlpool is continuously to dilute bacterial accumulation on ischemic or necrotic tissues.

Cotton pledgets between digits will prevent maceration of tissues. Bedside digital exercises of all the joints are recommended, and should be done throughout the entire waking day, and Buerger's exercises for the lower extremities are recommended four times daily. Narcotics are generally not utilized in uncomplicated cases after initial thawing; tranquilizers or aspirin will suffice for pain. In the past, sympathetic blockade, sympathectomy, anticoagulants, vasodilators, alcohol and enzymes have not proved particularly effective.

Over the years, many **surgical procedures** have been proposed for post-thaw care of the frostbite-injured extremity. The benefits of surgery should always be weighed against the possible injury to the regional vascular structures in the injured area. If possible, the surgical approach should improve the prognosis by relieving compartment pressure, increasing joint mobility, limiting infection, or increasing vascularity. The surgical procedures are: escharotomy; escharectomy; blister, bullae, and wound débridement; fasciotomy; arteriotomy; vascular wound repair; dermal graft procedures – Reverdin (Davis) pinch grafts, split thickness skin graft, split thickness skin graft (mesh), free full thickness skin graft, cutaneous pedicle flap graft, muscle/musculocutaneous vascular flap transfer, very early digital débridement with vascular cutaneous flaps; controlled subcutaneous balloon tissue expansion; guillotine or modified amputation; closed amputation (closed suction irrigation); closed or open reduction of fractures, dislocations; joint contracture releases, joint excision and replacement, joint fusion; soft tissue, web space releases; surgical regional sympathectomy; periarterial sympathectomy, microdigital sympathectomy; sinus tract and squamous cell carcinoma excision; and tissue compartment releases (carpal/tarsal tunnel syndrome) (Mills, 1983; Mills *et al.*, 1960 & 1961; Mills *et al.*, 1977; Franz *et al.*, 1978; King *et al.*, 1964).

In the past, the use of a **hyperbaric oxygen chamber**, single man unit, at two atmospheres of oxygen, has appeared beneficial in post-thaw frostbite therapy. Further evaluation of this adjunct method is planned as part of a study by the Department of High Latitude, University of Alaska.

In patients with apparently equal bilateral injury, results of sympathectomy within the first 24–48 hours have demonstrated that, while there is no further preservation of tissues, there is a decrease in pain, a marked decrease in edema, much less infection (superficial and deep), and early and more proximal tissue demarcation.

However, more recently, and still under evaluation, **sympathectomy and vasodilators** and sympathetic blockade have been determined to be of good effect following fasciotomy. The author suspects that the previous irregular results of these treatment methods often reflected an effort to perform effective sympathectomy, when the problem may have included regional vascular compartment pressure block or distal vascular thrombosis. The use of phenoxybenzamine hydrochloride has been particularly effective given in doses of 10 mg per day, and increased to 20–60 mg per day, depending upon effect and need. The drug is used for vasospasm, and appears to be an effective alpha adrenergic blocking agent. It is important that the patient be well hydrated after surgical or chemical sympathectomy. Pain varies with each individual and with the type of injury, the degree of edema and the presence or absence of infection. It is lessened by immediate physiotherapy, activity and a whirlpool bath. In severe cases of immersion injury, with edema, prior to fasciotomy, or with high-level extremity freezing, post-thaw, pain relief may be provided with continuous epidural block, for 24–48 hours, repeated if necessary. This is especially effective if accompanied by fasciotomy in severe cases with associated increased tissue compartment pressure.

In the last four years, especially in the pre-injury patient with a labile vasomotor peripheral vessel response, **biofeedback** has been utilized to increase the hand and foot circulation (**93**). This has been utilized as well in the post-thaw extremity freezing victim (Kappes and Mills, 1982).

93

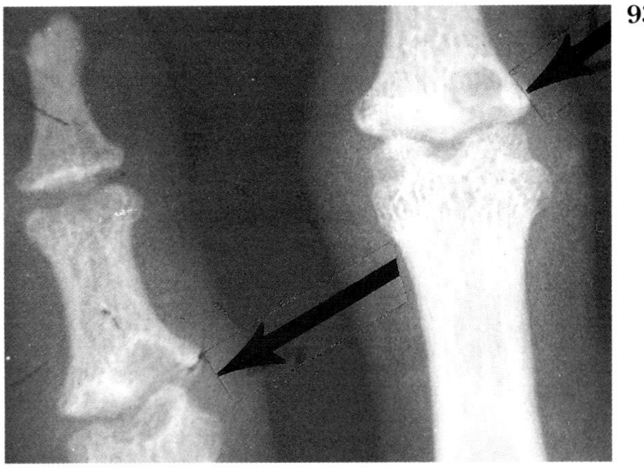

93 Demonstrated here are the periarticular and lithic destructive changes in the region of the proximal interphalangeal joint, secondary to severe cold injury. Thawing was by ice and snow, followed by rapid rewarming. The damage is presumed to be a result of: (i) destruction of the periarticular blood supply; (ii) direct injury to cartilage and bone.

Special aspects in treatment

Patients should be kept in a pleasant environment, not relegated to corners of the hospital because of odor or tissue necrosis. The diet should be high in protein and in calories, with vitamin supplements as needed. When considered necessary, anti-tetanus therapy is recommended, particularly a toxoid booster for those previously immunized. If amputation must be performed, a modified guillotine procedure at the lower level is recommended, with secondary closure to be carried out at a later date. Superficial or deep infection is often found in the extremity requiring guillotine amputation. Secondary closure after the amputation may be more successful when accompanied by closed suction-irrigation, with irrigation fluid (0.9% sodium chloride) flowing at 100ml per hour, with a flush of 50ml of the preferred antibiotic solution every hour. Dislocations and fractures pose interesting problems, and dislocation particularly should be reduced immediately after thawing. Traction, trauma, manipulation or open procedures are used rarely, and only with great care, in patients who had an extremity fracture prior to freezing. The fracture treatment should be conservative until the post-thaw edema is eliminated. It may be that a well-padded plastic mold is the best method of treatment until the edema ceases. If open reduction of fractures or dislocations is required, great care must be taken to avoid further vascular injury. Post-operatively, the operated part, in a plastic posterior mold, may still undergo whirlpool therapy and active digital exercises. The prognosis of this combined injury is poor, because added to the injury to the regional vascular supply from fracture trauma, is the superimposed freezing injury. It is here also that fasciotomy may be required to relieve the deep structure pressures. Fluids are encouraged, dehydration is to be avoided, and electrolyte balance should be maintained. Smoking is discouraged, but alcohol may be permitted.

The cartilaginous structures in children, particularly the epiphyseal plates and non-ossified carpal and tarsal bodies, are susceptible to cold insult and injury. Total necrosis of these organs is rapid and, at present, regardless of the methods of thawing and post-thaw care, the injury is apparently irreversible (**94** & **95**).

94

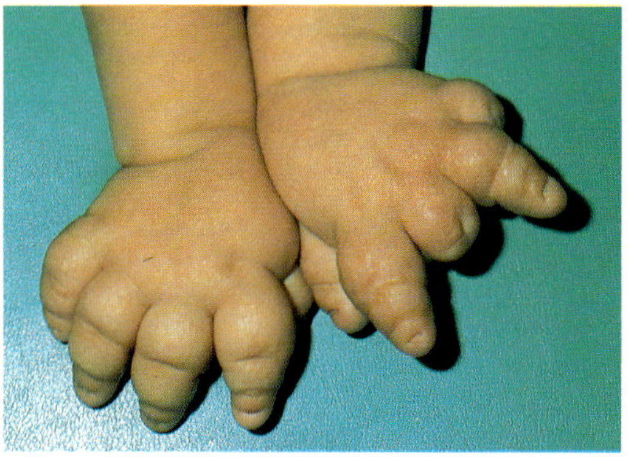

95

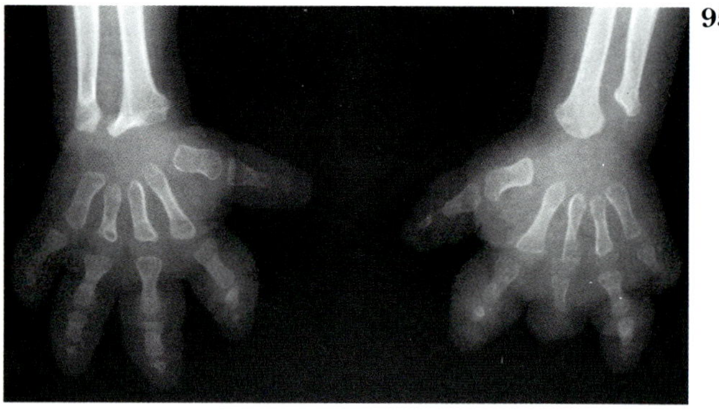

94 & 95 Severe freezing injury with severe hypothermia, in an infant left in the snow in midwinter. These pictures, taken at age five, demonstrate severe soft tissue changes, marked lymphedema and segmental phalangeal amputation. Furthermore, demonstrated in **95** is the complete destruction of the phalangeal epiphysis, as well as complete destruction and dissolution of the carpi bilaterally. Growth plate cartilage is very susceptible to cold, resulting in epiphyseal necrosis, digital deformity and shortening of the phalanges.

Concluding considerations

The above is a basic program, to which the physician may add any other therapy of choice, including low molecular weight dextran, vasodilators, anticoagulants, hypotensive agents, sympatholytic drugs and thrombolytic agents. Despite the best-intentioned treatment (regardless of thawing method, indicated drugs or surgical care), some results are unexplained disasters. These poor results may be due to extended depth and duration of freezing, repetitive freeze–thaw–refreeze injury, underlying circulatory deficit or to some other cause. The post-injury state may demonstrate: freezing injury with post-thaw vasoconstriction; freezing injury with arteriovenous capillary thrombosis; and severe cellular destruction as a result of the freezing.

Being aware of all the choices will help in the selection of the appropriate drug therapy. Anticoagulants (heparin), vasodilators (tolazoline hydrochloride) or hypotensive adrenergic blocking agents (guanethidine, reserpine), including sympatholytic drugs (phenoxybenzamine hydrochloride) may aid in the initial phase of care, especially in the absence of deep thrombosis. The plasma volume expander (low molecular weight dextran) used early, is thought to prevent, diminish or reverse red cell aggregation in the capillary tree. For deep occlusive thrombus formation, the use of thrombolytic enzymes (streptokinase and orokinase) is being evaluated. The risk of hemorrhage and lysis of fresh fibrin may limit the use of these drugs where there is associated trauma, especially head injury in which a cerebral vascular bleed may be of concern. The use of these drugs may then require special local and regional techniques. There is currently little help for the problem of severe or total cellular destruction.

In summary, the drugs used in frostbite injury care are: plasma volume expanders (low molecular weight dextran); vasodilating agents (tolazoline hydrochloride); hypotensive agents (guanethidine monosulfate, reserpine); hemorrheologic agents (oxpentifylline); calcium blocking agents (nifedipine); sympatholytic agents (phenoxybenzamine hydrochloride); anticoagulating agents (heparin); thrombolytic enzymes (streptokinase, tissue plasminogen activator – TPA); industrial solvent (dimethyl sulfoxide – DMSO); anti-inflammatory agents non-steroidal: acetylsalicylic acid, ibuprofen.

From the review of many cases, it has become apparent that the following points should be considered:

- Thawing should not be attempted where there is a danger of the injured part refreezing.
- Tissue damage has probably occurred at the level of the non-frozen–frozen interface in the process of survival, rescue, or early extremity handling during treatment.
- It would appear that the onset of hypothermia and frostbite may be the result of general overall dehydration and hypovolemia, resulting in further local distal tissue dehydration.
- It is again noted that the major disasters often occur when the individual has self-treated the extremities by an extreme of thawing temperature, using excessive heat, such as a camp fire, diesel exhaust, or oven heat. When there is inadequate hydration following injury, results are poor. If increased compartment tissue space pressures are not recognized and relieved, arterial access to the injured part and venous return are made difficult, if not impossible.
- In many cases, the points discussed above are of little use, as the thawing method is often out of the hands of the rescue workers and the attending physician, and a large number of patients brought to emergency centers or major hospital areas have already allowed thawing to occur, either deliberately by their own methods, or inadvertently while awaiting rescue or during rescue procedure. In these situations, post-thaw techniques and care assume the greatest importance.

Hypothermia

Bruce C. Paton

Hypothermia is not restricted to severe high alpine environments, to the arctic or even to winter. It can occur in summer, especially in wet and windy conditions, on land, on rivers, and after immersion in lakes and oceans, and develops whenever heat loss exceeds heat production and conservation. A core body temperature of less than 35°C signifies hypothermia (**96**).

96

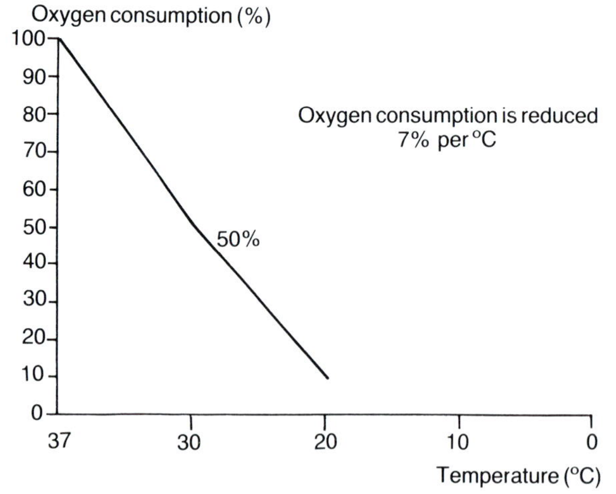

96 Oxygen consumption related to central temperature.

Classification

The usual classification is mild (35–32°C), moderate (32–28°C), and severe (below 28°C). A more practical classification is mild (35–32°C), and severe or deep (below 32°C) (**97**).

Mild hypothermia seldom results in permanent complications and, by itself, is virtually without mortality. However, respiratory problems, confusion, mental disorientation and difficulties with coordination do occur above 30°C. A danger of mild hypothermia is that body temperature may fall if rewarming does not occur. Occasionally, victims develop fatal cardiac arrhythmias, so there should never be complacency about the severity of the condition.

Below 28°C, many complications occur; deep, severe hypothermia is a dangerous and frequently fatal condition, which must be treated with urgency.

97

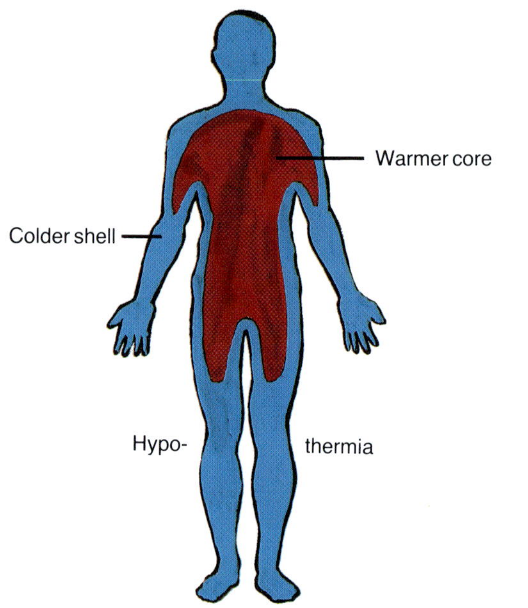

97 The central body temperature is the temperature of the core; the heart, lungs, kidneys and brain are preserved as long as possible from the cold.

Symptoms and signs

Symptoms

One problem in recognizing hypothermia is the lack of symptoms. Many people do not complain of discomfort and may protest that they feel fine, although they are not well and are becoming colder.

The first symptom is a feeling of increasing cold, accompanied by uncontrollable shivering. As cooling progresses, mental and physical processes deteriorate. Tiredness and exhaustion, sometimes out of proportion to the effort being exerted, cause the victim to think only of sleep, rest and lying down, although the conditions may be inappropriate. Sense of direction is impaired, and decision making is seriously affected. Hypothermia in a leader endangers the whole party. The chances of coma increase progressively below 32°C, although it is possible to remain conscious even at 26°C. Manual dexterity is impaired; buttons cannot be secured and zippers are left undone, because the effort to fasten them is too great. Gloves are lost and clothing left open. Cooling and frostbite result.

Occasionally, victims have a feeling of great heat, even to the extent that they remove clothes and are sometimes found naked or only partially clothed. Sexual assault may be wrongly implicated by investigating authorities.

Signs

General appearance. The appearance of the victim may vary between normality and apparent death. In mild cases, the face is normal, but in severe cases, the skin is pale and cyanotic and the face is puffy and swollen. The victim appears dead, with stiff limbs, imperceptible breathing and no palpable pulse.

Cardiovascular function. The physiological responses to cold are an attempt to maintain body temperature. Blood pressure, pulse rate and respiratory rate increase. Peripheral constriction reduces heat loss. As temperature falls, the decelerating action of cold exerts its influence: blood pressure falls, heart rate decreases, peripheral vascular resistance increases, and cardiac output diminishes. Were the patient euthermic, these changes would result in shock, but the demands for oxygen decrease in proportion to the fall in temperature, and cardiovascular function remains adequate for metabolic needs. Oxygen requirements fall by 7% per °C (50% of normal at 30°C; 10% at 20°C).

Below 30°C, cardiac arrhythmias are increasingly likely. Bradycardia is a normal response to cold, but may be extreme with heart rates of 20–30/min. Atrial fibrillation and flutter, heart block and, eventually, ventricular fibrillation or asystole are the most frequent disturbances of rhythm. Peripheral constriction increases vascular resistance. Some patients become vasodilated and the skin becomes bright pink.

Metabolic changes. Shivering is the first line of metabolic defense against cold, increasing heat production fivefold. The violence of shivering varies from person to person and can be very exhausting. Shivering stops at about 31°C.

Psychological and neurological changes. As the temperature falls, the victim becomes increasingly confused, disoriented and irritable with strange behaviour, before lapsing into unconsciousness. Nerve conduction is slowed, and the periphery becomes anesthetic. Injuries can occur without the person feeling pain. Movements are clumsy and poorly coordinated, because of slow nerve conduction and stiffening of the muscles. Reflexes disappear and the pupils become dilated and fixed.

Cerebral edema can develop. At high altitude it may be difficult to distinguish between changes resulting from hypothermia and those attributable to HACE (*see* pp 60–62, 65).

Respiratory function. While shivering is strong, hyperventilation occurs. At about 32°C, breathing becomes shallower and, in deep hypothermia, is hard to discern, with a respiratory rate of 4–6/min. Pneumonia and pulmonary edema are common complications and reduce oxygenating capacity.

Fluids and electrolytes. If hypothermia is combined with prolonged exposure, the victim will probably be seriously dehydrated with a low blood volume and little or no urinary output. There is also redistribution of fluid within the body, so that edema of the skin and lungs occurs in spite of general dehydration. Diuresis is the initial response of the kidneys to cold, further contributing to dehydration.

Electrolyte changes are variable. Serum potassium may be low or high, but is usually high in prolonged, severe hypothermia damage. High potassium levels contribute to cardiac arrhythmias. Potassium levels of 6–8 mmol/l have frequently been found, and even as high as 20 mmol/l. Such high levels are probably a sign that the victim is beyond resuscitation.

Glucose levels may be high or low. Pancreatic function is diminished and insulin (endogenous or exogenous) is not effective below 30°C. Victims exhausted by diminished adrenal function may require supplementary steroids. Acid–base changes are common and acidosis may be anticipated in all severe cases.

Other body systems. Gastric bleeding and pancreatitis are frequently found at autopsy. Coagulation is slow and defective, causing both thrombosis and bleeding.

Treatment

General principles

The following principles always apply: make a diagnosis accurately, with a deep reading thermometer; prevent further heat loss; rewarm appropriately (**98**); plan for evacuation.

98

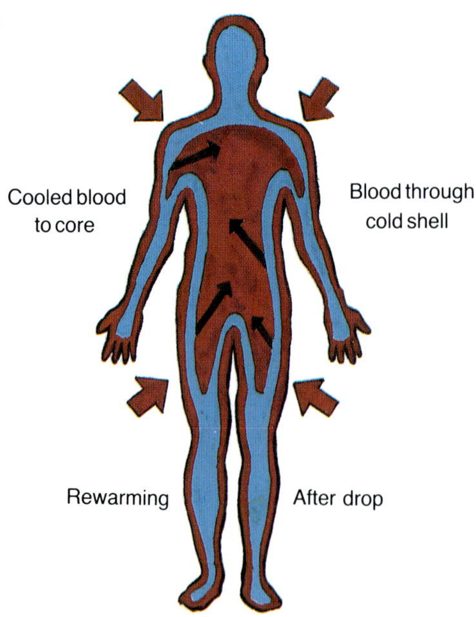

98 Inadequate reanimation may induce a circulation of cold blood from the shell to the core. This can lower the central temperature and produce heart failure. This mechanism is called 'after drop'.

Rewarming methods

Passive. Provide dry clothing and shelter. Wait for spontaneous rewarming.

Active. External: employ warmed blankets (electric or water-warmed), hot water bottles, circulating hot air, a hot water bath (42–45°C). The most efficient method is a hot water bath, capable of adding 2,520–8,400 kJ/h. These methods are suitable for all types of hypothermia, and require little special equipment. Most warm slowly.

Internal: give the patient hot food and drink; employ gastric, thoracic and peritoneal lavage, warmed inhaled air (max. temp. 45°C), peritoneal and hemo-dialysis, femorofemoral extracorporeal circulation (**99**). These methods are invasive and should be used only in severe cases, when other methods cannot provide enough heat. Special equipment and skills are necessary. Extracorporeal circulation is the most effective method for warming and resuscitating an asystolic patient.

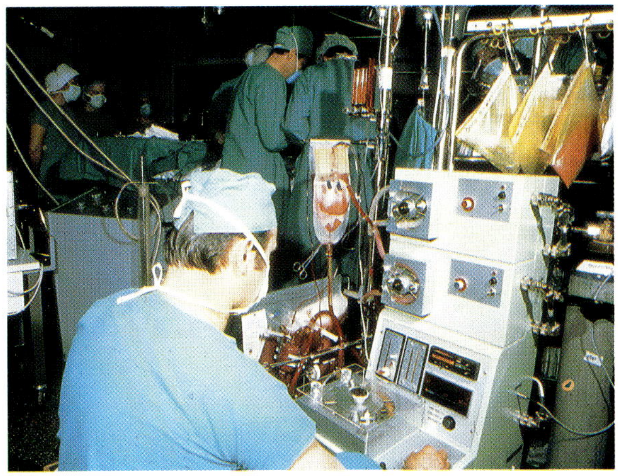

99

99 Extracorporeal circulation is the best rewarming method for deep accidental hypothermia. *Courtesy of P. Segantini.*

Mild hypothermia

Spontaneous rewarming is possible in all cases, and therefore the methods chosen need only prevent further heat loss and provide extra heat by simple means.

Severe hypothermia

In the field, consider the time available for rescue before starting active rewarming. Slow methods are better than fast ones; most practical, available methods limit the speed of rewarming. Monitor vital signs; do not perform cardiopulmonary resuscitation (CPR) unless it is certain that cardiac arrest has occurred.

In hospital, full intensive care is essential for the control of cardiac, respiratory, acid–base and fluid problems. Choose a rewarming method suitable for the patient's condition and the hospital facilities.

Special problems

Ventricular fibrillation is potentially induced by handling, intubation, nasotracheal suction or insertion of intravascular catheters. Acidosis, respiratory alkalosis and high serum potassium levels accentuate the tendency to fibrillation. Bretylium tosylate is the most effective antifibrillatory drug at low temperatures, and can be used prophylactically.

Resuscitation

If cardiac arrest is confirmed, resuscitate and maintain resuscitation until death is beyond doubt. If the victim is very cold, and cardiac action is uncertain, do not perform CPR until a diagnosis of cardiac arrest has been made. Waiting may be safer. The chances of restoring cardiac action under 30°C are poor. Rewarm the patient before attempting to defibrillate. Invasive methods are necessary to rewarm in the absence of natural circulation. CPR does not provide sufficient circulation to rewarm.

Results

Recovery from hypothermia by itself, is usually complete and without complication.

Mortality rates vary with age, general health, severity of hypothermia, duration of exposure, cardiac status and correctibility of electrolyte and acid–base problems. They range from 0–95%. Mild hypothermia should be without mortality. Severe hypothermia always has the potential to be lethal. The young survive better than the old, and the healthy better than the sick. Brief hypothermia is safer than prolonged.

Hypothermia in an avalanche, a case report

Adam P. Fischer, Frank Stumpe and Jacques Vallotton

In fine weather, in February 1990, a group of skiers were caught out by an avalanche. One person was covered by the snow. The victim relates her impressions:

I set off with six other people on a cross-country ski run, along an established route in the western Alps. Each of us had an Autophon VS68-type transceiver. As it was quite warm, I was only wearing a T-shirt and ski trousers, and I was carrying a mountain backpack. Although the weather was fine, the risk of avalanche was quite high, and on several occasions during the climb we heard crackling noises in the snowfield. The surface snow was powdery and thin, and covered an older, grainier layer of snow.

At 13.30 (H 0), just after we had started on the final slope at a height of 1,900 m, the mass of snow suddenly detached itself 50 m above us, over a width of 200 m. At that moment, I was in the middle of a kick turn, and I undoubtedly panicked and fell. Before I could get up, the avalanche reached me and carried me along for about 80 m. I was immediately covered with snow and unable to withstand the force of the avalanche. When the avalanche stopped, I was completely buried, unable to move, lying on my side, with my right arm stretched out in front of me. At first I panicked, but then I thought it would be better to remain calm and to conserve my energy. Thanks to my backpack, which had tipped over my head, my face was protected and my mouth was free. After a dozen deep breaths, I lost consciousness.

The companions of this 46-year-old woman immediately began to search for her, and one of them went to alert the air rescue services (REGA). The skier was quickly located, thanks to the transceiver she was carrying.

At 14.30 (H + 1), the victim's friends succeeded in releasing her head, which was buried under 2 m of snow. She made some faint breathing movements and gave a few moans.

At 15.10 (H + 1h 40), the helicopter arrived at the scene of the accident with rescue equipment. The patient had already been pulled out of the avalanche and was unconscious. An intravenous catheter was inserted, and, after endotracheal intubation, ventilation and cardiac massage were begun. The patient's pupils were midsize but not reactive, and the tympanic temperature was 15.4°C (however, the snow which filled the external auditory canal made this figure unreliable). A portable cardiac monitor first revealed an asystole, and then a ventricular fibrillation which became apparent during external cardiac massage. The patient was placed in the helicopter and, during the 12-minute flight to the hospital, cardiac massage and ventilation were continued.

At 16.10 (H + 2h 40), the helicopter landed at the hospital's heliport (CHUV, Lausanne). The patient was livid, and appeared to be dead, with fixed, dilated pupils and acute pulmonary edema. The rectal temperature was 19.6°C. The cardiac monitor revealed a slow ventricular fibrillation, but attempts to reduce this were unsuccessful. An arterial catheter was inserted into the left femoral artery. Systolic pressure attained through cardiac massage barely reached 4.0–6.7kPa, at a rate of 20/min. Ventilation was provided with six insufflations per minute. The pulmonary edema was severe and required frequent endotracheal aspiration. Blood gases (H + 3h) pH = 7.22; $P\text{O}_2$ = 3.8kPa; $P\text{CO}_2$ = 5.7kPa; BE (base excess) = − 11.6mmol/l; HCO_3 = 20.1mmol/l (*see also* **Table 3**).

At 17.45 (H + 4h 15), active rewarming was started using extracorporeal circulation (ECC) with a partial femorofemoral bypass. At an esophageal temperature of 24.6°C, the pupils became constricted. The heart started beating spontaneously at 26.4°C, initially in a nodal rhythm, which subsequently reverted to a sinus rhythm at 90/min.

At 04.30 the following morning (H + 15h), the patient was extubated and remembered everything up to the point when she lost consciousness in the snow. Subsequently, the patient was found to have a parasternal contusion, some pain in the tenth left rib and a small fracture of the sheath of the right shoulder blade. Nine days later, she left hospital.

Table 3 Laboratory data.

	H + 3	H + 7.5	D + 1	D + 2
K (mmol/l)	4.5	2.5	3.6	3.9
Lactic acid (mmol/l)	8.92		2.06	
Na (mmol/l)	143	147	147	145
CK (UI/l)	138	853	5470	9030
CK MB (UI/l)	100	95	78	135
Amylase (UI/l)	96		248	199
PT (prothrombin time) (%)	65	50	65	80
PTT (partial thrombin time) (s)	36	38	46	28
Fibrinogen (mmol/l)	1.6	1.2	1.9	2.6

H = hour of accident; D = day of accident

Hypothermia in a crevasse, a case report

Frédéric Dubas and Jürg Pfenninger

In July 1983, a 13-year-old boy went out for lunch at Langfluh Restaurant (2846 m above sea level) above Saas-Fee, Switzerland. After lunch, at 13.45, he went alone for a walk upon the glacier that surrounds the restaurant and fell into a crevasse. He plunged about 10 m and became stuck where the crevasse narrowed. His father found him five minutes later and called for help.

After 15 minutes, the rescue team arrived by helicopter, with a trained rescue doctor and all the equipment necessary to dig someone out of an ice coffin. The rescue was extremely difficult as the boy was wedged very deeply in the crevasse. He became unconscious at 15.45 and was only pulled out at 17.00. At this time he appeared to be dead, with cardiorespiratory arrest and dilated, unreactive pupils. He received slow cardiorespiratory resuscitation and massive protection from the cold. He was transported to the University Hospital of Berne, the nearest hospital equipped with ECC (extracorporeal circulation), arriving at 17.45.

Upon arrival the boy's core temperature was 17.1°C; Biology: Hb 12.4 g/dl, Ht 0.38, platelets 162.000, Na = 141 mmol/l, K = 3.1 mmol/l, glucose = 12.1, mmol/l PT (prothrombin time) = 33%. Blood gases were pH = 7.23, PCO_2 = 5.6 kPa, HCO_3 = 19.3 mmol/l, BE (base excess) = − 8.7 mmol/l. EEG showed diffuse slow waves with delta and theta waves, mainly on the temporo-occipital right side. There were no epileptic potentials. X ray of the lungs showed diffuse edema on the left more than the right, and slight mediastinal emphysema. All other tests, including a CT scan of the skull, were normal.

Treatment and evolution

Active rewarming was conducted by ECC with a femorofemoral partial bypass for an hour. Heart defibrillation was performed when the core temperature reached 30°C, after which the heart rhythm became normal. The patient was admitted to the Intensive Care Unit at 22.00, with a core temperature of 32.6°C. He reached a core temperature of 37°C after 14 hours, and a peripheral temperature of 37°C after 24 hours (**100**). Intubation was maintained for three days with a maximal PEEP (positive end-expiratory pressure) of 8. The FIO_2 was progressively reduced from 1 to 0.4. He received phenobarbital, calcium and cefatrexyl. Evolution was easy without any complications. The pulmonary edema, a usual appearance after hypothermia, probably due to the hypoxic damage of the small blood vessels, was progressively reducible.

The neurological evolution was encouraging. After initial depression (abnormal crying on the first day after extubation), his personality quickly became normal. A short-lived diplopia persisted for one day. His left arm showed a slight plexus brachialis syndrome with peripheral lesions of the Nn. medianus and ulnaris, probably due more to a compression during the accident than to hypothermia. After eight days in hospital the boy was able to return to school and suffered no complications from the accident.

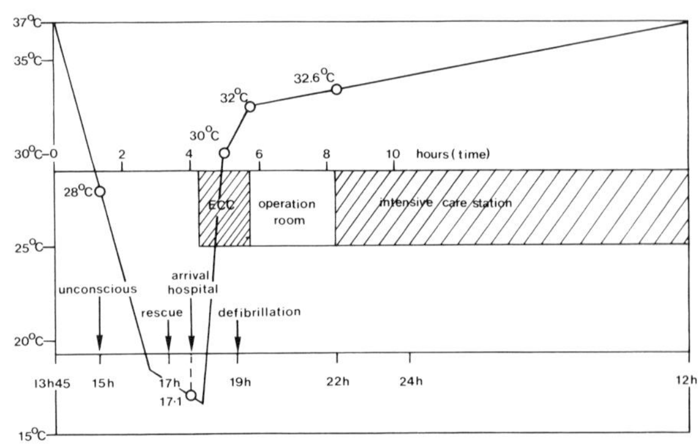

100

100 Synopsis of the case report.

Ice and snow

Personal experience of an avalanche

Georgy Vianin

In February 1984, a fellow guide and ski instructor and I decided to take a group of skiers on a route which stretched from the summit of the Corne de Sorebois towards the resort of Grimentz, via Tzirouc, in the Zinal height. However, unknown to us this route, which was open that day and which is very popular in good conditions even though no ski runs are marked out, had become unstable.

We set off. I was at the front with my colleague, who I asked to stay back and oversee our clients through the first tricky stage of the run, which had to be done by sideslipping. Meanwhile, I continued across a slope, to reach the top of a large valley, namely the Tzirouc Pass. I waited on a shelf about 10 m below the cornice for the others to join me one by one. Suddenly, without any noise, the ground began to move; I realized what was happening immediately and tried to get away by skiing on but, unfortunately, the whole slope in front of me was also coming down, and I found myself in the heart of the avalanche.

Everything happened very quickly. After skiing a few dozen meters, I struck a moving block of snow and lost my skis. It was here that my years of experience almost certainly saved my life: the thought struck me that I had to release my ski poles immediately.

From then on, I was caught up in a giddy descent of several hundred meters. I was completely lucid and knew exactly what was happening to me. All the theories which I had known for years, such as making swimming movements and protecting my face with my hands, were impossible to carry out; I was powerless. At times I was crushed under great piles of snow, and at others I was whirled around with the feeling of being torn limb from limb.

It is impossible to describe exactly what went through my head during this fall, and it was not until several months later that I was able to collect my thoughts. It was not so much the thought of death that frightened me, but the idea of the physical suffering and slow suffocation, and I remember calculating the time it would take the rescuers to release me.

When the avalanche stopped, I miraculously found myself on the surface of the snow. I remember learning that everyone was safe and sound, but, probably due to the shock, the rest of the day has remained a blank in my memory.

Avalanches and their mechanisms

André Roch

101

101 Ice avalanche on the Nevado Illimani, Bolivian Andes.

In 90% of avalanche accidents, it is the victim who triggers the avalanche (**101**) which carries him away. To prevent this happening, it is important to understand the mechanisms which set avalanches in motion.

Snow consists of ice crystals and air. Snow crystals form in the atmosphere, and fall on the ground where they continually change shape. The mechanical characteristics of the fallen snow depend on the type of crystals and on the crystals' mutual cohesion.

The weather transforms the snow crystals, both as they fall and on the ground. The wind rolls, breaks and packs them into compact layers called wind-slabs. The snow is blown away on the exposed slopes and accumulates in the sheltered areas. A rise in the air temperature penetrates the snow slowly and decreases its cohesiveness, beginning on the surface. The sun's radiation modifies the surface, changing the crystals to grains or melting them. A frost layer covered by a new snowfall is dangerous, because the frost may act as a lubricant.

The crystals falling on the ground are shaped as stars, needles, plates and other forms (**102**). The hexagonal stars entangle their branches and form light powder snow. Newly-fallen snow is very compressible; the bottom layers of deep snow are therefore packed together by the weight of the overlying snow. The snow on the surface remains light, and, with time, changes slowly into fine, and later coarse, grains. A layer that has remained a long time at the surface loses compressibility and resists packing under the weight of further snowfalls. It remains weak within the snow blanket, whereas the already packed inferior layers remain compressible and become more and more compact.

The change in the snow crystals from star to grain may occur at a constant temperature, and is then called isothermal metamorphism or destructive metamorphism (**103**). The fine branches of the stars disappear, as water vapor escapes from the points of the needles and settles in the hollows to form grains.

The temperature varies within the snow blanket. At ground level, the snow has the same temperature as the ground; if the ground is frozen at the beginning of winter, its temperature returns to 0°C, due to the earth's heat and the insulating effect of the snow which contains a large amount of air. The snow at ground level will be close to 0°C. On the other hand, at the snow's surface, the temperature follows the average air temperature, which depends on the altitude and the period of winter. During cold spells, the temperature difference between the ground layer and the surface layer is great. The relatively warm air at the bottom of the snow blanket is lighter than the cold air, and therefore rises towards the surface. This warm air is saturated with water vapor. When it cools, it can no longer hold as much vapor, and this generally condenses on appropriately oriented crystals, which grow into prisms or cups; other crystals tend to disappear. This is called temperature gradient metamorphism or constructive metamorphism (**104**). It causes a new structuring of the layers of snow, which become a fragile structure, susceptible to collapse and likely to set avalanches in motion. A very cold spell activates gradient metamorphism, and a temperature rise decreases the snow's cohesiveness.

The stratified snow blanket is formed of compressible, compact layers and strata of coarse crystals, which are less compressible and remain brittle. This stratification varies with time and location, according to the frequency and abundance of snowfalls, the temperature, and the effects of terrain on snow loading.

102

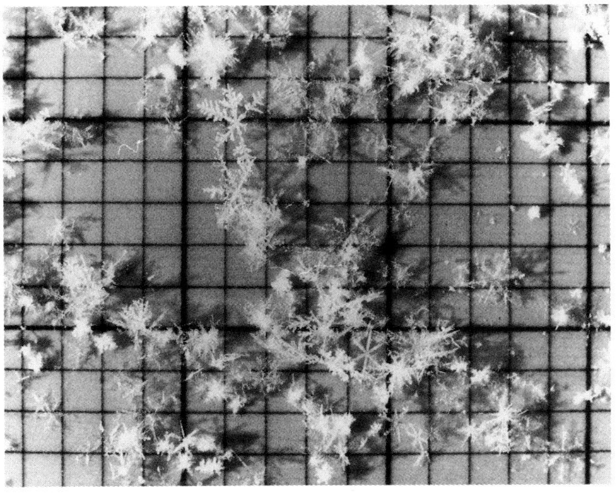

102 Snow crystals.

103

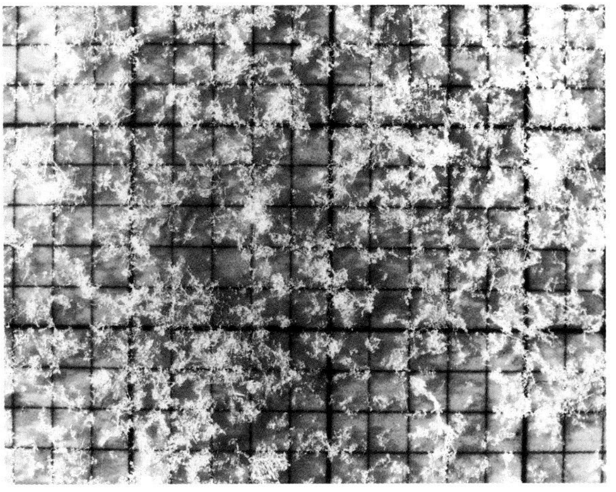

103 Destructive metamorphism.

104

104 Constructive metamorphism.

The release of avalanches is a mechanical problem. We can consider, for example, a snow blanket on a slope. At each depth of the snow blanket, the resistance of the snow to shearing forces can be compared to the force along the hill's slope of the weight of the snow above. A stability coefficient can be determined (**105**); it is the ratio of the resistance over the stress. As long as the resistance is greater than the stress, the snow is in equilibrium. We must therefore consider the various elements which either increase the stress or decrease the resistance.

A snowfall can be so abundant that its weight exceeds the resistance of the most fragile underlying layer, and releases an avalanche. Also, the new snowfall causes an increase in the temperature of the older, underlying layers, which decreases their resistance.

The structural modifications due to gradient metamorphism induce a slow decrease in the resistance of fragile layers, which can break and release an avalanche. These two mechanisms of induction set off so-called spontaneous avalanches, as opposed to shock- or accident-triggered avalanches, which can be intrinsic or external (**106**).

105

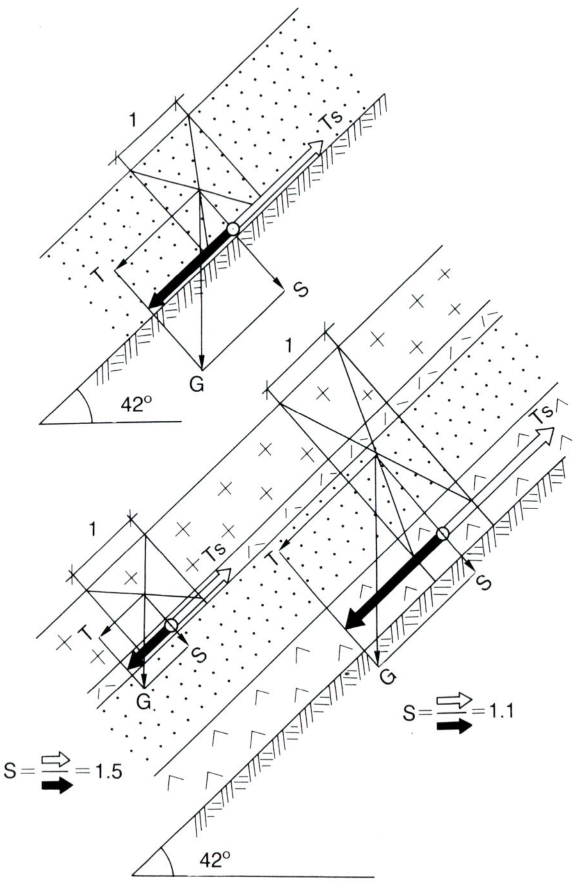

106

105 Measure of the stability index at different levels of the snowpack.

G = Gravitation.
S=T/Ts = Stability index.
T = Mathematic resultant of the weight related to the slope.
Ts = Mathematic resultant of the friction related to the slope.

Spontaneous release of an avalanche can be due to: (a) a new snowfall (increase of T); or (b) the weakening of breakable layers, caused by the temperature gradient metamorphism by a warming up of layers (decrease of Ts).

106 Study of the different layers of the snowpack and evaluation of their cohesion. *Courtesy of J. Michelet.*

On a convex slope, the snow flows more rapidly on the steep part. The layers are stretched and can break. The more compact the stratum is, the greater the shock of the rupture. This can loosen the bond between the crystals in the brittle layers, which then act as a lubricant for the motion of the overlying layers. This is an example of an avalanche triggered by an intrinsic shock. Shocks may also be external, for example the impact of a falling snow ledge, snow falling from trees or from a wall, or even the weight of a skier.

As one can see, the mechanism is complex, and it is difficult to determine the cause of an avalanche's release. The best way of estimating the danger is to analyze the factors which cause avalanches:

- A **snowfall,** which overloads and warms the old layers.
- The **wind,** which flattens the snow and heaps it up in irregular depths; thick layers reduce the stability of the snow blanket.
- A **warming up** of the layers, which reduces their resistance; the water resulting from the thawing weakens the bonds between the crystals, and acts as a lubricant for granule movement.
- The **stratification:** the presence of brittle internal layers, which are not adherent to an underlying hard layer or to ice (**107 & 108**).

The skier or mountaineer may evaluate the situation by means of several tests. For example, he may plant a ski pole or dig a hole to look for a layer of low resistance. Alternatively, he may look for lubricating layers by freeing a rectangular surface and jumping on it to see if the snow will slide. Lastly, in order not to be taken by surprise by an avalanche, one should always listen to the weather forecast, to the avalanche warnings, and observe nature. However, experience proves that, in spite of all precautions, one will always be fooled by avalanches.

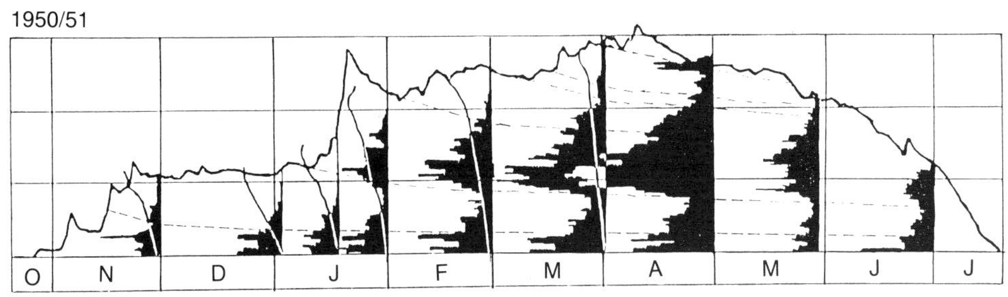

107

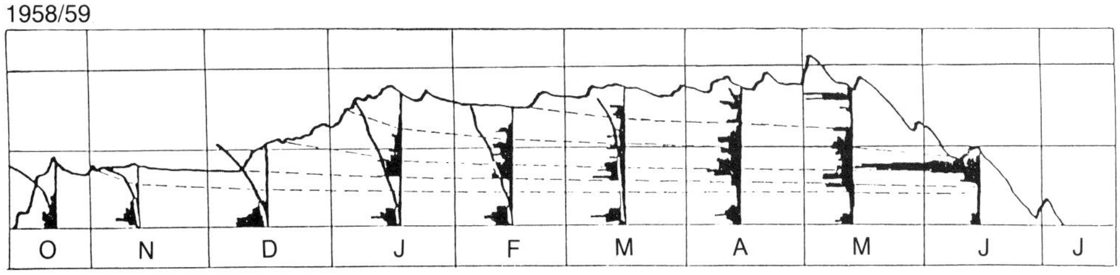

108

107 & 108 Evolution of the hardness of the snowpack during two different winters at the Weissfluhjoch, Davos, Switzerland.

In 1950–51, the layer of the November snowfall stayed at the surface for two months, lost its compressibility and remained weak during the whole winter. In Switzerland, there were 98 people killed, 80 in villages and houses, and 1,301 cases of damage.

In 1958–59, there was little snow, and the snowpack was weak throughout the winter. There were 15 casualties, 3 in houses, and 18 cases of damage.

Avalanches: prevention and rescue

Frédéric Dubas, Rémy Henzelin and Jacques Michelet

While there is a risk of avalanche on any snow-covered slope, there are factors, namely certain weather and snow conditions, which can make its occurrence more likely. Anyone venturing into the high mountains must take account of these factors, and be aware of the risks.

Skiing away from regular ski runs has become very popular in recent years. It satisfies the skier's need for risk, space and freedom, and begins where the ski lifts end, outside the safety of established skiing areas. When conditions are critical, there are always skiers ready to ignore the prohibitive signs and flashing lights indicating a danger of avalanches, and ready to try a run without being fully aware of the dangers they are facing.

While many mountain skiers are well informed and know that theirs is a difficult and often a dangerous environment, there remain various visitors to the mountains who should be made aware of the dangers to which they are exposing themselves. There are an increasing number and variety of information sources available to these skiers, such as specialized magazines on mountains, skiing, snow conditions, avalanches and rescue techniques, and films and television programs. In addition, there are practical and theoretical training courses, which provide the necessary information about avalanches and what to do in the event of an accident.

In recent years, considerable progress has been made in both short- and medium-term weather forecasting, and this plays a vital role in skiing excursions. In dangerous weather conditions, skiers aware of the forecast are able to change their route or, if necessary, their final destination.

During winter, the snow and avalanche study institutes analyze the snow blanket and forecast the avalanche risks. Their reports inform the skiers of conditions in the Alps and help them plan their excursions accordingly. In addition, those responsible for the safe running of the ski lifts are constantly observing and monitoring the snow blanket, especially at spots where avalanches often begin. The deliberate triggering of avalanches in controlled circumstances also contributes to the skiers' safety.

If setting out on an excursion in the mountains, the skier will find the following advice very useful:

- Find out about the condition of your planned ski route and about any tricky parts of it, and take the weather of the previous few days into account. Take great care over difficult terrain or in bad weather conditions, and rope yourselves together in areas with crevasses.
- Most accidents occur when a group chooses the shortest, most pleasant route, instead of a safer but possibly more difficult one.
- When skiing in areas where there is always some risk of avalanche, you must know the region well, or be accompanied by a guide or by experienced companions.
- If you set out on your own (not recommended), inform someone of your route and of your probable time of return. There must always be somebody who knows about your plans and who, if necessary, will know where to look for you.
- Do not assume tracks outside marked ski runs are safe; they may have been made in different snow conditions, during a rescue, or to deliberately trigger an avalanche.
- If the risks of avalanche are high, be strong-willed and give up your plans.

Safe conduct on mountain excursions

When going up a mountain, the skier should always follow its crests and ridges, which are exposed to the wind, rather than its valleys, which are generally filled with snow. Rocks, flat areas of land, and trees are reliable points of reference when planning a route. Skis should be carried, if necessary (**109**). When approaching a ridge, it must be remembered that a cornice is nearly always indicative of a wind plate (snow loading) (**110**), and the area should be avoided, or crossed with the usual safety precautions to avoid overload. Stay on the windward side of the ridge, and be aware of any cracks in the snow blanket. Stay well to windward of these. Skiing a slope with wide crosses and kick turns should be avoided as this cuts up all the snow in the valley, and the kick turns leave piles of snow at certain points, which the snow blanket cannot support. Therefore, it is sometimes best to pass through on foot, one behind the other, following the line of the slope.

When crossing a dangerous area, the skiers should be spaced out so that they go over it one at a time. While a skier is crossing a slope, the others should watch both him and the ground constantly, so that they can warn him if necessary. If the worst comes to the worst, i.e. if an avalanche begins, the others should try to fix his point of disappearance. These precautions must be observed until the last skier has crossed the area, as the safety of a slope does not increase with the number of crossings. When crossing dangerous slopes, the skier must release the straps of his skis and take off the wrist-straps of his ski poles. If the skier is carrying a pack, the waist-belt should be unfastened, and the pack carried by one strap while the hazardous slope is being crossed. After he has crossed, he should find a safe place to stop, but must remain alert and ready to act, should an avalanche occur.

109

109 Dangerous slopes should always be crossed as high as possible.

110

110 Particular attention must be paid to cornices.

When coming back down the mountain, the skiers should only leave their shelter (chalet, hut, bivouac, etc.) if conditions are favourable. The route down must be carefully planned, so that any slopes with a risk of avalanche may be avoided. In addition, the skiers should decide on a plan of action and an escape route to follow in the event of an avalanche. By keeping to the sides of the slopes when descending, the skiers will be able to avoid an avalanche more easily. The order of descent should be decided and the skiers' speed must be controlled, to reduce the risk of falling. If a skier does fall, he must try to get up again immediately. Skiers should descend by making sharp turns or tight parallel curves (**111**). Crossing the slope must be avoided unless absolutely necessary, and the descent should be adapted to suit the ability of the weakest skier. The ski pole wrist-straps, the straps of the skis and the waist-belts of the skiers' backpacks must be released. Their mouths should be protected, with scarves, for example.

111

111 During the descent, skiers should cross dangerous areas one by one, and should avoid making wide curves.

What to do in an avalanche

Avalanches are natural phenomena of such intensity that it is difficult not to panic if caught in one. However, it is vital to remain calm and to act according to the following instructions.

If the avalanche begins under your feet, try to ski away to the side and, with luck, you may be able to get out of the avalanche area. If you are knocked over, try to remain on the surface of the snow by letting go of your ski poles and, if possible, your skis and back pack. It is better to lose your equipment than to be dragged down with it. If you are near trees, bushes, rocks or other obstacles, hang on firmly; this is only possible at the beginning of an avalanche, before it builds up speed. Try to stay on the surface of the snow and to get to the edges of the avalanche by making strong swimming movements.

Keep your mouth closed. If you lose your sense of direction, roll yourself up into a ball. Sometimes it is best to protect your face. Before the mass of snow stops moving, put your arms across your chest and your hands in front of your face, and relax. This will create an air pocket which will enable you to breathe, for when the avalanche finally stops, the snow solidifies very quickly and prevents any movement. At the moment the snow stops moving, attempt to determine which way is up by spitting. If possible, thrust an arm or leg in this direction. This will possibly aid in your recovery, as the portion of your anatomy closest to the surface will be found first.

If you are conscious, listen for the calls of those who escaped the avalanche and give short sharp cries as soon as your companions are nearby. Snow is very absorbent and sound passes through it with difficulty. Remain calm; the most important thing is to breathe and to believe help is on its way.

If you escape the avalanche or witness it, you should first keep watching for those who have been carried away. Next, mark the places where they have disappeared and begin to search around these points. Call for assistance if you are carrying a radio beacon. Look carefully for any item which is sticking out or lying on top of the snow. Do not carry away any objects found, but mark their position. Call and listen; a shout may be heard.

Attempt to determine where the person may be buried by observing the form of the terrain, any possible signs and the quantity of snow. Look especially at shelves, as these frequently mark obstacles where the snow has been held back. If your party is large, organize yourselves so that those who escaped the avalanche can begin searching immediately, either by using their transceivers or by probing the snow with ski poles or with the heels of skis. Appoint someone to coordinate the searches, and to act as a lookout for further avalanches. Continue searching with conviction until help arrives and do not, under any circumstances, let fatigue get the better of you.

The transceiver

The transceiver (**112**) (*see* p. 207) is by no means a guarantee against every risk, but it can help in a rescue after an avalanche. The fact that its use is still not widespread is probably due to its inability to provide a total guarantee of survival, to its expense, and to its maintenance requirements (the battery in particular). In addition, training in the use of a transceiver is essential.

The transceiver must be both a radio transmitter and a radio receiver, and in the future, there should be standardization of the many available models and, in particular, of the frequency used. Sets that are solely either transmitters or receivers can be used only in established skiing areas, which are supervised from a resort and allow the rescuers to be at the scene of the accident within a few minutes (**113**).

112

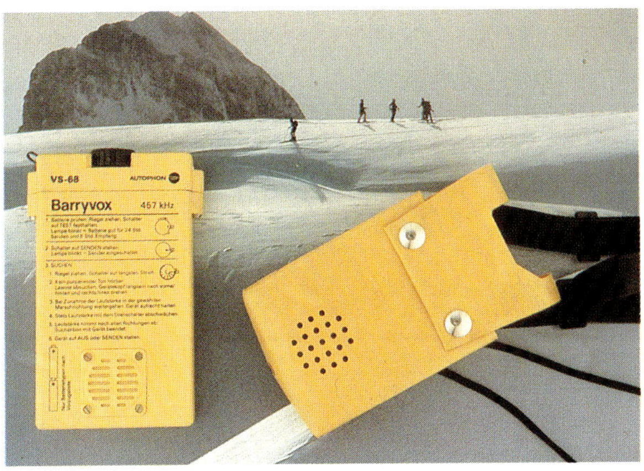

113

112 If its user is well trained, the transceiver allows an immediate and effective search for avalanche victims to be made.

113 Devices that are solely either receivers or transmitters are not well adapted for mountain rescue. Pictured here is one of the heavy, cumbersome pieces of equipment that the rescue team must carry to the scene of the accident, when a receiver or transmitter is used. *Courtesy of Swiss Air Rescue.*

Avalanche rescue

Statistics

Over the last ten years, our knowledge of the medical problems experienced by avalanche victims has greatly increased. However, the prognosis is still gloomy, and in most cases, those pulled from the snow are already dead. Exact statistics are difficult to establish both for the total number of cases, and for the proportion of known survivors to the overall number of those who escape the avalanche. In fact, a large percentage of accidents not requiring rescue resources and involving no fatalities, are not reported to the authorities and thus escape the records.

These factors must be taken into account when interpreting the statistics, which therefore merely indicate trends. The best-known study is that which resulted in the definition of the famous De Quervain Mortality Curve (*see* **15**, p. 22). This curve is derived from the study of 2,611 cases of avalanche victims, buried at an average depth of 1.06m, between 1945 and 1979. A study of such cases occurring between 1979 and 1989 has confirmed this curve. On analysis, the curve shows that one hour after the accident, 30–40% of the victims may still be alive, and that this percentage drops to less than 10% four hours after the accident. Therefore, the victims must be located as quickly as possible and freed from the snow without delay (**114**).

114

114 The freeing of an avalanche victim. A medical evaluation can be performed in 10 seconds. Cardiorespiratory resuscitation must be undertaken as soon as possible. *Courtesy of Swiss Air Rescue.*

Pathology

In avalanche accidents, specific injuries must be distinguished from non-specific injuries.

Non-specific injuries

Non-specific injuries constitute any pathology resulting from being dragged along, crushed and pulled at by the moving mass of snow, and being thrown against obstacles such as rocks and tree trunks. Non-specific injuries include fractures and depressions of the skull, chest and abdomen, and injuries to the spine. They are the cause of 20% of the initial deaths and of a large number of the subsequent deaths.

Specific injuries

Asphyxia is by far the most common cause of death, representing 70–80% of the fatalities. It may be either rapid or slow.

Rapid asphyxia can occur by the following means:

(i) By compression: where there is an avalanche of wet snow, the weight of the mass of snow may reach or even exceed 800 kg/m^3. When this mass is set in motion, it exerts a pressure of several tonnes per square meter. After breathing out heavily, the victim may find that breathing in is impossible due to the pressure exerted on the thorax.

(ii) By obstruction of the upper air passages: if the mouth becomes filled with powdery or wet snow, immediate asphyxia results. Irritation of the larynx by the small crystals of powdery snow may give rise to a reflex laryngospasm. Snow rarely gets into the lungs, but if it does the effects are the same as in drowning.

(iii) By injury to the respiratory system: fracture of the chest, hemopneumothorax, pharyngotracheal tearing by barotraumatism (see below).

Slow asphyxia can occur by the following means:

(i) Through a disturbance of the diffusion of air through the mass of snow; this may produce a reduction in the oxygen concentration around the injured person.

(ii) Hypothermia or other traumatic cerebral injuries may reduce the sensitivity of the respiratory center.

(iii) Traumatic injuries can progressively hinder the respiratory system, for example hemopneumothorax and compression of the thorax.

The blast caused by a large avalanche is responsible for much of the damage that occurs every winter to the natural environment (e.g. to mountain forests) and to constructions such as bridges, roads and railways. It also causes **barotrauma** – injury to the exposed organs of the body, such as the eardrums, the pharynx and the pulmonary parenchyma.

Hypothermia is discussed at length in a separate section (*see* pp. 92–98). When buried by an avalanche, it is rare that the victim survives long enough to suffer from hypothermia; he is generally killed by asphyxia, before hypothermia has a chance to set in. The speed at which hypothermia occurs varies, depending both on external factors and on the individual.

The main external factors are the type of snow and the clothing. Damp snow wets and rapidly reduces the insulating power of the victim's clothes; he will therefore come into close contact with the cold mass more quickly. Insulating and waterproof clothing maintains the victim's core body temperature longer.

The main factors concerning the individual are the thickness of the layer of fat, the degree of exhaustion, and age. An evenly distributed layer of fat is extremely effective in maintaining the victim's body temperature. To fight against hypothermia, intact glycogen reserves are essential, as they are consumed to produce heat through shivering. Therefore, if the body has already exhausted its reserves (for example, by prolonged efforts) it will be much more vulnerable to cold.

Hypothermia occurs more easily in elderly persons, chiefly for metabolic reasons, and in children, because of their smaller body volume.

Frostbite may also occur, but it is rare and mostly superficial.

Rescuing the victim

The chances of survival in an avalanche are greatly increased by the skier's knowledge of, and ability to take, the correct course of action, and by the training, skill, availability and speed of the rescue services. In avalanche rescue, an important distinction must be made between immediate rescue through the efforts of the victim's companions, and organized, professional rescue.

Immediate rescue

Most people who have the good fortune to survive an avalanche have been saved by companions present at the time of the accident (**115**). Visual and auditory searches can begin immediately, combined with attempts to locate the victim by probing with ski poles and the backs of skis, or by using a transceiver in the probable area of burial. Searches with transceivers will be successful only if those employing them are fully acquainted with their use. Unfortunately, this is not always the case, and the lack of training and practice is paid for by the death of the victim.

115

115 An immediate rescue can be undertaken by the companions of the victim, by listening and looking, using transceivers and performing improvised sounding with ski poles.

Organized rescue

The main qualities of such a service are its efficiency, availability and speed. The most essential tool in avalanche rescue today is the helicopter, with its pilot and flight technician. Few rescue operations are conducted without it.

Despite the various searching methods now available, avalanche dogs and their handlers remain the most effective tools in the search for people buried in an avalanche (**116**).

Every organized and effective rescue operation involves the apointment of someone to take charge, capable of requesting the necessary resources immediately, and of directing and coordinating the rescue. This task is generally entrusted to a guide, to the head of a rescue unit or to a professional rescuer. Doctors must also be present (if possible, one for each person buried), together with the rest of the rescue team, bearing the essential equipment, such as probes (**117**), shovels and transceivers.

In some cases, the presence of too many rescuers, helicopters, dogs, machines, radios, etc., at the scene of the accident can make it virtually impossible to locate buried victims. Depending on the weather and the nature of the terrain, exhaust gases can sometimes pollute the cone of the avalanche. Snow blown about by the helicopter rotor can block pores and cracks in the snow surface, through which the victims' calls may be heard. The noise of helicopters, radios and other devices can greatly disturb the avalanche dogs in their searches. The person in charge must therefore decide which tools are to be used in the search and when these should be replaced by others. Someone competent must also be appointed to communicate by radio with the rescuers, together with someone to note the topography and the progress of the searches. This allows the person in charge of the operation to make considered, coordinated decisions, in full possession of the facts.

16

117

116 Avalanche dogs are still the alpinist's best friend in avalanche rescues.

117 A sounding team, equipped with long aluminum sticks allowing sounding to a depth of 4 m.

Medical attention for the avalanche victim

(i) **A brief examination as soon as the head can be reached:**

- Consciousness.
- Breathing.
- Circulation.
- Hemorrhage.

(ii) **If the patient is fully conscious and the cardiorespiratory system is functioning normally:**

- Continue with the rescue operation.
- Provide maximum protection against the cold; give the victim hot, sweet drinks.
- Prevent the victim from moving, in order to avoid cold peripheral blood flowing back towards the center of the body; this causes the temperature of the heart to drop and can result in serious rhythm problems (after drop).
- Measure the core body temperature using the epitympanic thermometer (**118**).

118

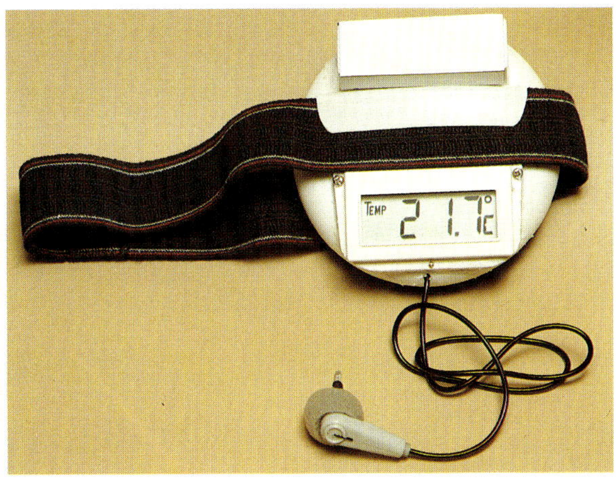

118 A good approximation of the central temperature can be obtained at the scene of the accident through epitympanic measuring, using the thermometer pictured here (*see also* p. 209).

- Quickly take the patient to a medical center, experienced in problems connected with hypothermia, and where an anesthetist is available 24 hours a day, as well as an intensive care unit.
- There must be strict monitoring during the journey to hospital, since a hypothermal patient may experience a cardiorespiratory arrest at any time.

(iii) **If the patient suffers a cardiorespiratory arrest** (the arrest may be due to severe hypothermia, to asphyxia, or to a traumatism which, unfortunately, is much more common):

- Begin resuscitation measures immediately: ventilation (intubation), external cardiac massage.
- Measure the core body temperature using a epitympanic thermometer; if it reads 28–30°C or higher, the apparent death is not due to hypothermia.
- Make sure you know the circumstances of the avalanche and details of the discovery of the victim, so you can specify: the estimated time of the cardiac arrest as, if it occurred more than one hour previously, the chances of resuscitation are very slim; whether an air cavity was present around the victim's face, enabling him to breathe, or whether, on the contrary, his mouth was full of snow – a factor suggesting a very bad prognosis.
- Other non-scientific considerations may be of importance and should be noted by the rescuers; these include the general impression of the victim's condition, whether the body is livid or has discolored patches, and the temperature of the thorax, measured by putting a hand under the victim's shirt. However, it should be remembered that these methods are empirical and not accurate.

If serious hypothermia is suspected (core body temperature lower than or equal to 28–30°C), and circumstances suggest resuscitation is possible, the patient must be taken immediately to a hospital with a cardiac surgery team available around the clock, and which can perform rapid rewarming by means of extracorporeal circulation.

Other methods of rewarming have been described for patients suffering from deep hypothermia (peritoneal dialysis, dialysis by means of a Swan–Ganz catheter, slow warming on a bed of ice, warm gastric lavages, etc.). These methods have produced good results when used correctly. However, the authors prefer rapid rewarming via extracorporeal circulation, by means of a partial femorofemoral bypass. When performed correctly, this method of rewarming proves to be the most reliable and the most effective.

Rescue in crevasses

Frédéric Dubas, Rémy Henzelin and Jacques Michelet

Glaciers

The ice which comprises the glacier is in a constant state of change, expanding and contracting as it adapts itself to the shape of the land (**119**). Generally, the central part of a glacier slips more quickly than its edges, and crevasses are formed from the resulting tension. The most common are those which open up between the moving block of ice and the part of the glacier that remains attached to the rock face. These crevasses are known as 'rimayes' or 'bergschrunds' and are perpendicular to the slope of the glacier.

There are also crevasses that lie across and along the glacier, forming on uneven, bumpy stretches of land. The former open up in the direction of the glacier's movement, and the latter open up along its axis. Cross-shaped or star-shaped crevasses are created by rocks lifting up the ice. There are also marginal crevasses which form on the edge of glaciers, and are due to the difference in speed of movement between the central block of ice and the edge of the glacier. This type of crevasse is also formed by the heat which the rocks radiate.

If part of the glacier suddenly falls away, crevasses called 'séracs' form in all directions (**120**). There is then a risk that the remaining wall of ice will collapse at any time, under the thrust of the glacier.

119

120

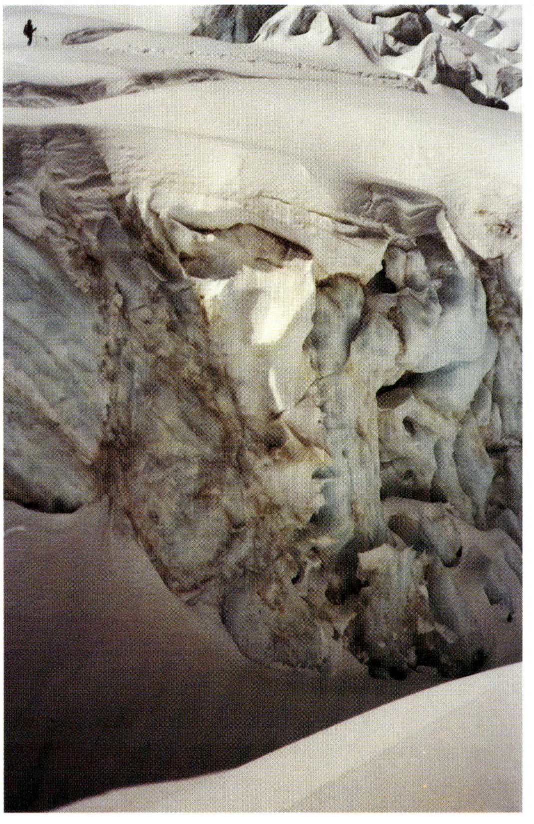

119 A glacier is constantly in motion and may be compared with a slow-running river (Weisshorn, 4,505 m).

120 Crevasses can be very deep and wide.

The prevention of accidents and safe conduct

Generally, the safest time to cross a glacier is towards the end of winter and in the spring, when the crevasses are covered by a fairly substantial layer of snow. The greatest care must be taken when the glacier has only a thin covering of snow, or if it is exposed to the wind.

The strength of a snow bridge can best be assessed by observing it from the side (**121**). Any visible shadows, sunken areas or cracks should be sounded out with an ice axe or a ski pole.

During an outing on foot, the participants must be roped together at all times when crossing a snow-covered glacier. Skiers, however, are generally only roped together during the ascent. Otherwise, when skiing back down the mountain, one skier falling into a crevasse might drag all the others with him. In principle, roping together is recommended only in foggy conditions where the skier might lose the trail, or on a particularly dangerous glacier with several crevasses and weak snow bridges. When roped together on an excursion, the rope must be kept taut and no attempt should be made to cross a glacier, unless equipped with a full-body harness.

121 The strength of a snow bridge depends on various factors, such as the heaviness of the snow fall in the previous winter, the temperature and the wind.

Rescuing a victim

A fall into a crevasse can be fatal. Rescue should be attempted only by someone skilled in improvised crevasse rescue techniques. If such a person is not present, efforts should be made to reassure the victim, the crevasse should be clearly marked and the alarm raised as quickly as possible.

Crevasse rescue is an area in which a great deal of progress has yet to be made. As with avalanches, each fall into a crevasse presents the rescuers with a grueling race against time, as they attempt to release the victim before he succumbs to the cold. In addition, the victim often has injuries, caused by the impact of his fall against the icy walls of the crevasse. These need to be considered by the rescue team, together with the difficulties of mounting a suspension system, or of rescuing a victim buried in snow or ice.

On falling into a crevasse, the weight of the victim's body can leave him wedged between the icy walls, compressing his thorax and resulting in severe breathing difficulties. Where this occurs, the victim's body heat melts the ice and he sinks deeper and deeper into the crevasse. The only means of rescue then is by clearing a passage through the ice with an ice axe or a pneumatic drill and compressor (**122**).

122 Specially developed pneumatic drills may be useful for digging out ice-trapped victims. *Courtesy of F. Bertin.*

123

124

123 The tripod is indubitably the most helpful device in crevasse rescue.

124 Improvised rescue without a tripod is also possible, but may be difficult.

The tripod (**123** & **124**) is without doubt the best means of extracting a seriously injured person from a crevasse with the least possible discomfort. In a straightforward rescue operation, i.e. where there is no need to cut the ice, three professional rescuers suffice. However, seven rescuers must be present at any operation involving the use of a pneumatic drill (fitted with ice pins and a compressor), which is specially adapted for the high mountains and may be carried by helicopter (**125**). The drill and compressor allow the removal of several cubic meters of ice per hour, provided that there is room for the falling ice to escape down the crevasse, and that it is not necessary to bring it all up to the surface, in what is an extremely time-consuming process.

125

125 An organized crevasse rescue operation requires a team of at least three to seven rescuers. *Courtesy of F. Bertin.*

When all that is needed is to release a victim stuck to the ice, thawing liquid (developed by France's Platoon of High-Mountain Gendarmes) is both rapid and effective (**126**). This liquid, kept in pressurized containers, is sprayed onto the ice, causing it to melt rapidly.

The 'Fernod–Ked' splint, developed in the United States, has also proven very effective in rescue operations in narrow crevasses, or in crevasses which are not easily accessible. The splint prevents any movement to the spine, and so is especially useful if there is a suspected injury to the spinal column.

126

126 Defrosting liquid has been developed by the Chamonix rescue team and is sometimes used to disengage victims. *Courtesy of J. Foray.*

Medical problems associated with falls into crevasses

A fall into a crevasse can result in a great variety of common injuries. Among the more specific are:

Hypothermia: (*see* pp. 92–98) in crevasses, the conditions greatly favor the development of hypothermia. The victim generally has enough air not to suffer from asphyxia. In addition, he may be unable to move, and be covered by snow which fell into the crevasse with him. The conditions are then identical to those found in an avalanche.

Frostbite: (*see* pp. 73–91) close contact with the icy walls of the crevasse often causes frostbite; this is mainly superficial.

Injuries from ice axes and crampons: climbers who fall into crevasses are usually equipped with crampons and ice axes. Moreover, they are generally roped together and sometimes fall several at a time. Such falls can result in serious injuries (**127**), such as perforation of the chest or abdomen, and injuries to the face or to the eyes. The rescuers carry the same equipment and may, while conducting a rescue, injure a victim or one another. However, these injuries are rarely serious.

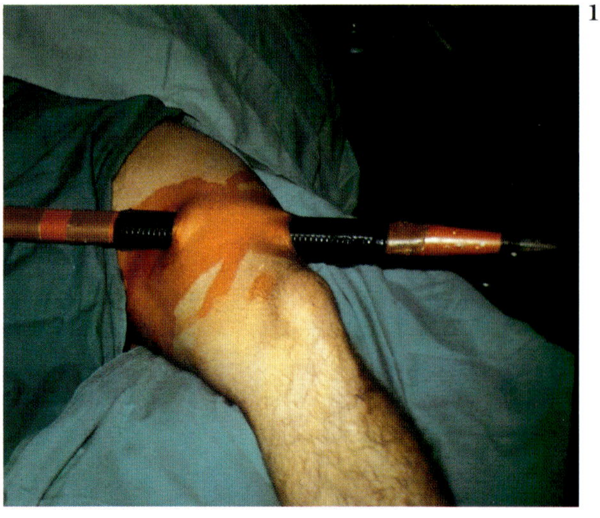

127

127 Ice axes may be dangerous; pictured here is a transfixant wound that occurred during a crevasse fall. *Courtesy of J. Foray.*

Lightning

Olivier Baptiste, Alain Girer and Jacques Foray

Lightning is a visible electrical discharge, usually associated with a thunderstorm, and caused when the increasing electrical charge within a cloud, classically cumulonimbus, overcomes the insulative capacity of the surrounding air. This discharge is known as the flash.

Mountaineers risk injury by lightning strike, owing to their proportionately greater exposure to conditions that favor thunderstorms: rapidly changing climatic conditions and exposed terrain. In order to properly evaluate the risks and to act intelligently to avoid them, it is necessary to understand the mechanisms of lightning.

The climatic conditions that presage a thunderstorm involve either a rapidly moving cold front or intense heating of the ground surface, with the resultant convectional uplift of air. Cumulonimbus clouds usually form when the ascending air current rises past the condensation level. The formation of such clouds can be taken as a sign that polarization is occurring. In effect, the cloud becomes a huge battery, and discharge will always follow.

Terrain plays a role in the location or probability of lightning strike. Summits, ridges, and isolated trees should be viewed with caution during a thunderstorm. Similarly, caves, hollows, gulleys, overhangs, and water courses should be avoided.

The lightning flash consists of a line of highly ionized air molecules along which electricity briefly flows. The initial discharge, of low luminosity, travels from cloud to ground, but is quickly followed by a highly luminous return stroke which comes to meet it from the ground along the same ionized channel. This is the main visible flash, capable of transmitting a current of 10,000 A and travelling at 0.1 the speed of light. Lightning can manifest in several forms, for example ball, chain, forked, ribbon, sheet or streak. Possibly the most common manifestation to the mountaineer is the eerie presence of St. Elmo's fire, a form of ball lightning that floats above any metal object or rock prominence.

The bi-directionality of lightning strikes has practical consequences to the mountaineer. A strike can occur in any of the following ways:

- A direct strike.
- A lateral or reflected strike.
- The ground transmission of a strike.

Pathology

The nervous system

- A loss of consciousness occurs in 80% of cases, and is always accompanied by retrograde amnesia.
- Brain lesions occur rarely.
- When a lightning strike hits the skull, hemiplagia, hemiparesis and intracerebral bruising may be observed. If there is damage at the medullar level, paraplegia may be seen.
- Injuries to the cranial nerves have also been observed, especially to the auditory and facial nerves.
- Peripheral lesions, like the 'keraunoparalysis' of Charcot, with vasomotor troubles may simulate acute ischemia, but this regresses within 24 hours.

The cardiovascular system

- Unlike neurological lesions, which are common, heart attacks have been observed only when the electric current has crossed the heart.
- The heart symptoms are either ischemic or take the form of a disturbance in the rhythm. The evolution may vary, depending on the localization and seriousness of the injury.

The skin

- Lightning may cause cutaneous lesions called 'Lichtenberg drawings' (**128**), which resemble the pattern of lightning on the skin. However, this condition is badly defined and inconsistent.
- Burns of varying severity are seen frequently, with the most serious ones located around the entry and the exit points of the lightning. They may be aggravated if the person struck is wearing metallic objects (**129–131**).

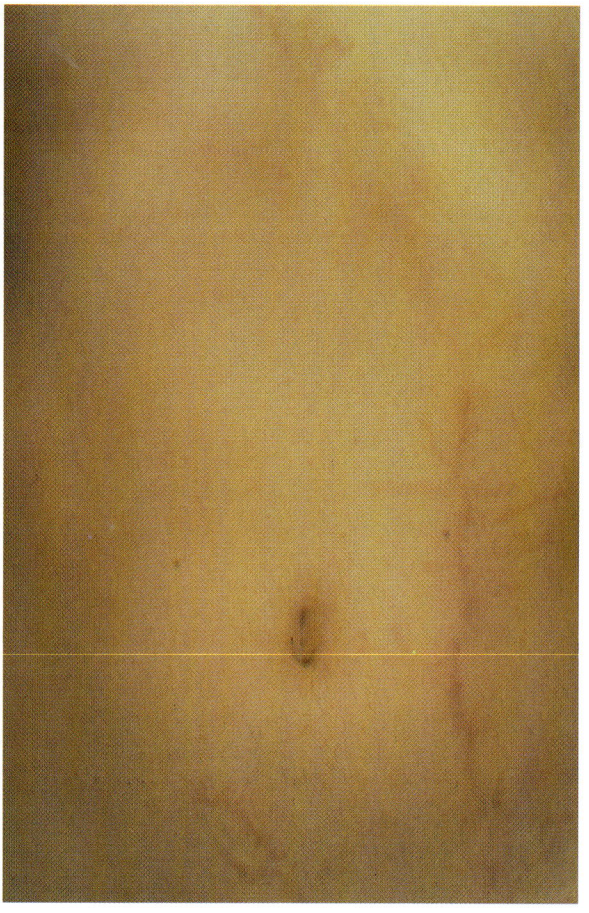

128

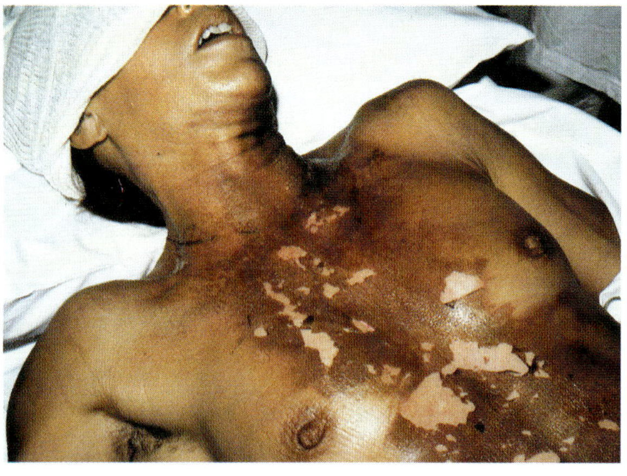

129

128 Lichtenberg drawings on the abdominal skin.

129 Thoracic lesions with sequelae caused by a melted gold chain.

130 Entry point of lightning on a thigh.

131 Exit point on the leg.

130

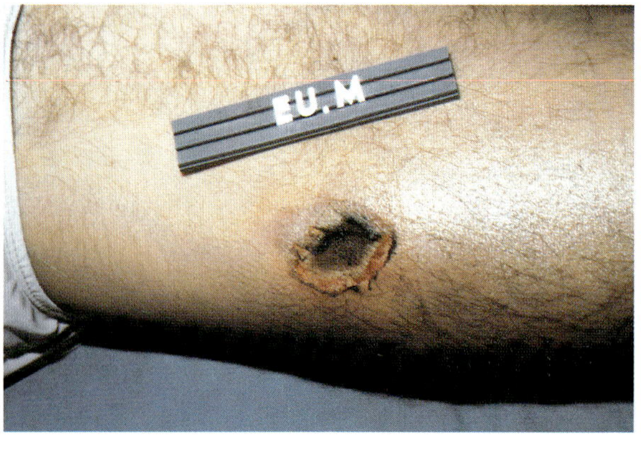

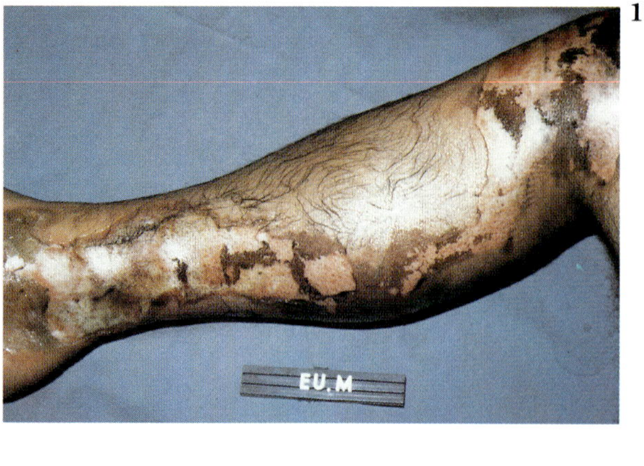

131

Other areas

- Other lesions are rare. Pulmonary disorders with a respiratory distress syndrome are found occasionally. Ocular, digestive and psychiatric disorders are sometimes present.
- In cases where the victim is pregnant, a maternal death has never been observed by the authors, but in 50% of such cases, the fetus dies.

In Chamonix, the authors have treated 30 people hit by lightning. The observed pathology is quite different when frostbite and hypothermia are associated conditions (**132**), and these can sometimes play a protective role. Mortality due to falls following a lightning strike is more significant.

The surgical and medical treatment is usually symptomatic. Burns, hypothermia and the often associated frostbite receive specific treatment.

132

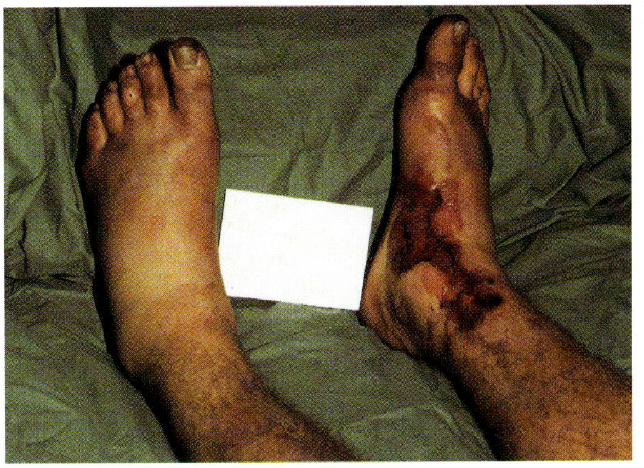

132 Association of frostbite lesions on both feet and lightning burns on the right foot.

Prevention

The best form of prevention remains the decision not to set out for a climb when weather conditions are unstable. In addition, there are some simple rules that can be followed to avoid danger during a thunderstorm. Lightning can strike directly or indirectly. One should therefore move away from all summits or ridges, where the ascending flashes of lightning are often fatal. Similarly, one should avoid taking shelter beneath a metallic object or wearing metallic headgear, although the metal itself will not attract more lightning than the body; on the contrary, it is likely to deflect the current from its surface. However, direct contact between the metal and the skin can produce burns.

Ground currents are very dangerous, especially in humid zones and areas with running water. The climber should assume a crouching position, and avoid cracks, overhanging rocks and any direct contact with the rock face. The best position to avoid ground currents is therefore crouching on a rope or a rucksack, at a distance from the rock face, with the hands on the knees, and the head bent forward. The lightning current can spread along rope, so abseiling/rappeling is to be avoided.

133

PART II

TRAUMATOLOGY AND MOUNTAIN SPORTS

133 Free climbers: a new generation of mountaineers. *Courtesy of C. Rémy.*

Hiking and Trekking

Injuries related to hiking and trekking

Albino Lanzetta and Remo Seratoni

In recent years, more and more people have taken up hiking and trekking, both professionally and as a hobby. This has led to an increase in the number of injuries related to these forms of exercise. Below is a concise description of the causes, diagnosis and treatment of the most frequent trekking injuries, involving the most commonly injured part of the body – the foot (**134**). Simple advice is given both to beginners and to experienced climbers on how to prevent the occurrence of such injuries.

People who go on trekking tours may be divided into two groups: first, those in excellent physical condition who go hiking and trekking regularly; and second, those who climb mountains less often, as a hobby, and who therefore cannot take as much stress as the professionals. For both groups, good health and good physical condition are necessary. In the first group, the most frequent pathologies are: **stress fractures**, especially at the metatarsal level; **tendinitis** and **insertionitis**, especially of the Achilles tendon or of the plantar fascia; and **muscular** or **tendinous ruptures**. **Sprains** or **muscular strain**, and **blisters** caused by unsuitable shoes should also be included. This distinction should not, of course, be considered absolute, but only indicative.

134

134 A selection of good trekking boots.

Symptoms and clinical observations

Blisters

Blisters occur frequently and are sometimes so painful that they make further walking nearly impossible. They are generally caused by new footwear that has not been broken in. The shoes are often either too tight or too large, and are worn with unsuitable socks, causing local friction and, consequently, large blisters (**135**). The most commonly affected areas are the toes, the heel and the perimalleolar regions. The foot is hyperemic around the injury, with very painful blisters filled with serum and/or blood.

135

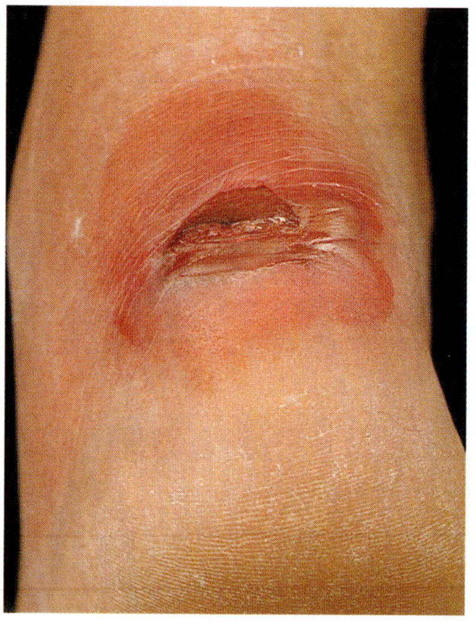

135 Infection is a common complication of blisters.

Insertionitis and tendinitis

Insertionitis and tendinitis generally occur in the Achilles tendon and in the plantar fascia. Achilles tendinitis is, by definition, a painful inflammation which may be accompanied by swelling. In the acute phase, it is only the sheath that is involved. In the chronic phase, a nodule of mucoid degeneration with longitudinal fissuring may develop in the body of the tendon. Tendinitis may be caused by

136

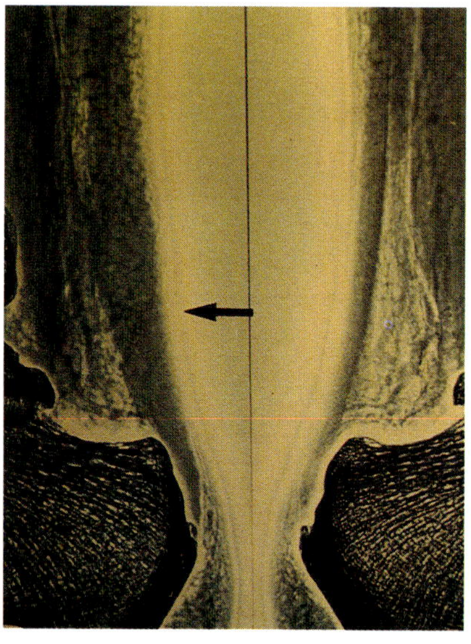

136 Xeroradiography shows inflammation and enlargement of the Achilles tendon.

137

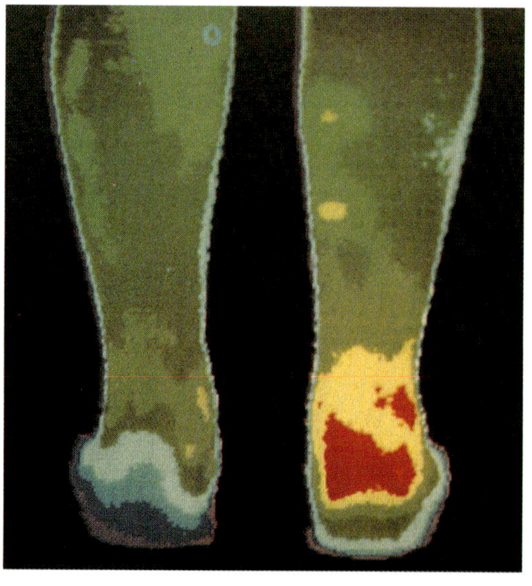

137 Thermography shows inflammation of the Achilles tendon.

long walks uphill, especially when carrying weight on the shoulders (i.e. climbing) and wearing footwear with extremely rigid soles. Running downhill places additional stress on the tendon. Unsuitable shoes, with an insufficiently padded heel support, and shoes with overly soft heels or wedge-shaped inserts may favor this injury. Tibia vara, functional equine or hollow feet, and a slight sprain of the gastrocnemius and soleus muscles can also favor tendinitis.

Repeated stress may cause small cumulative lacerations, which finally induce an inflammatory process that is clinically detectable. Pain is usually localized in a segment 4–5 cm above the tendinous insertion, but may spread to involve the whole tendon down to the calcaneal tuberosity. In addition, a posterior calcaneal bursitis may be found, as a result of tendinitis or as its cause. Therefore, the onset of painful symptoms affecting the Achilles tendon should lead to a precise diagnosis, assisted by clinical examination and by non-invasive tests, such as low-voltage X rays, xeroradiography (**136**), thermography (**137**) and ultrasonography.

Plantar fasciitis or 'calcaneal spur syndrome'

Plantar fasciitis is the most frequent cause of heel pain. It is caused by overload, and implies an inflammatory process at the attachment of the plantar fascia to the calcaneum and, in serious cases, a calcaneal spur. Frequent microtrauma to the sole of the foot or to the calcaneum is the cause of the lesion, which occurs chiefly when walking downhill on very bumpy paths, wearing footwear that does not protect the plantar region properly (very light gym shoes, for example). In the acute phase, the patient may feel pain when he starts walking in the morning and, as often happens with other stress injuries, he may feel it diminish during physical activity and return later once the exercise is over. Additional problems may be caused by an alteration in the plantar articular biomechanics, such as increased calcaneal varus or valgus and adductus of the forefoot. Local palpation causes intense pain at the attachment of the plantar fascia to the heel. An X ray may show a calcaneal spur (**138**). Thermography produces a thermic diagram with diffused or circumscribed hyperthermia.

138

138 X ray demonstrating a calcaneal spur.

Sprains or musculotendinous ruptures

These lesions are relatively common, especially in the gastrocnemius muscles. They often occur in subjects who are insufficiently fit. The patient generally complains of a shooting pain or sharp cramps, followed by functional restitution. Such injuries often arise when undertaking difficult or very long climbs. Again, unsuitable footwear or the presence of altered biomechanics in the foot are both conditions favoring the onset of this pathology. Muscular ruptures, particularly of the gastrocnemius muscles, mainly result from violent falls, such as when the climber is using crampons and is overloaded.

Generally, the rupture of the Achilles tendon is

the final step of a progressive degenerative process, involving the entire musculotendinous apparatus. The rupture often appears in subjects who, until it occurs, are asymptomatic or only feel a slight dull pain in the calf. When moving the ankle in a sudden (or sometimes even a smooth) way, the patient feels a sharp pain in the posterior part of the calf. Walking immediately becomes difficult or even impossible. Physical examination shows a rupture of the tendon, which on palpation feels like a depression. Later, the tendon swells up and responds positively to Simmond's test (squeezing the affected calf of the prone patient does not produce plantar flexion of the foot) (**139**). Low-voltage X rays can reveal discontinuity of the damaged tendon, and hemorrhage. Ultrasonography shows a typical acoustic signal at the point of rupture. Thermography indicates a diffuse hyperthermic gradient.

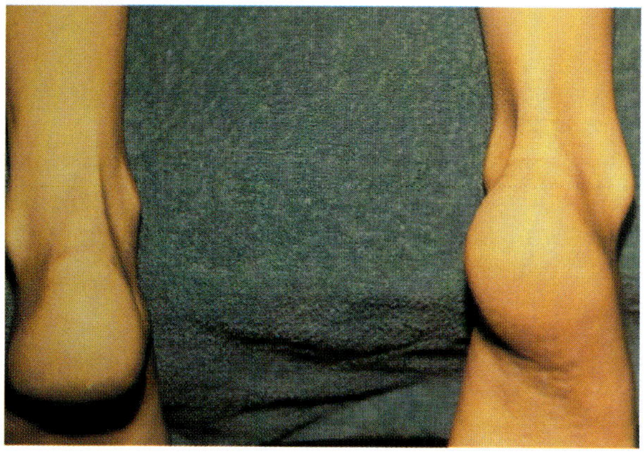

139

139 Inability to accomplish symmetric plantar flexion in a subcutaneous Achilles tendon rupture.

Stress fractures

Stress fractures are discrete interruptions of the bone and are due to repetitive functional stresses, which gradually overcome the bone's resistance to fracture. In the foot, they mainly involve the second (**140**), third and fourth metatarsal diaphyses. These fractures are often caused by long walks uphill, on hard surfaces such as rock, or by repeated efforts being made on tiptoe. The patient complains of recurrent metatarsalgia or pain under overloading conditions. Pain is also aroused by pressure, but swelling is rarely present. A precise diagnosis is usually possible by radiographic examination of the foot. Sometimes, there is no radiological evidence for a stress fracture; in these cases, a technetium scintigram is useful.

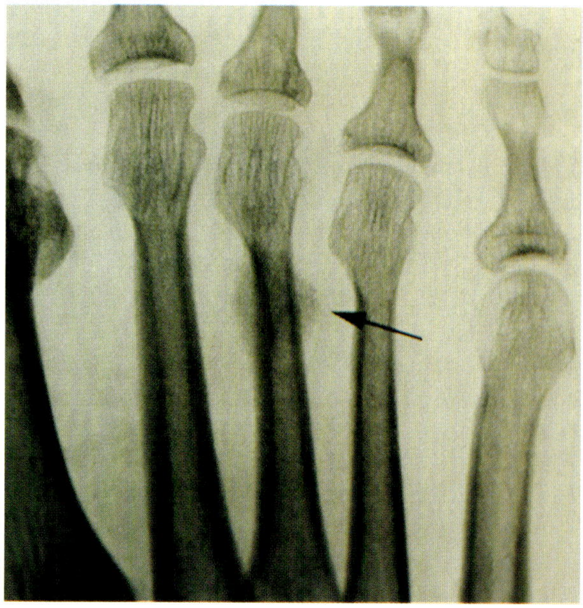

140

140 Radiograph of a stress fracture of the second metatarsal bone of the right foot.

Treatment

Blisters

If blisters are filled with serum and/or blood, they should be disinfected and cleaned locally to avoid bacterial infection. Clothing and shoes should be checked to ensure they are comfortable and properly shaped.

Achilles tendinitis

The treatment of the acute phase of Achilles tendinitis is rest, local ice packs, iontophoresis and ultrasound therapy. Shoes must have a flexible sole, well-shaped heel padding and a wedge-shaped heel support at least 15 mm thick (to be inserted in the footwear) and rigid lateral support for the heel. The heel support helps to relax the tendon while walking. Specific orthopedic appliances are required when the articular biomechanics of the foot are altered. If the tendinitis is in its chronic phase, it may be necessary to immobilize the foot in a slight equinus position in a cast for four weeks. In some cases, surgery is the only way to restore a satisfactory anatomy to the tendon and avoid rupture.

Plantar fasciitis and posterior calcaneal bursitis

The therapy prescribed for posterior calcaneal bursitis is similar to that for tendinitis in its acute phase. Local infiltrations of corticosteroids are not recommended, as these may cause secondary rupture of the tendon. Heel padding and ice packs may not be adequate in the treatment of plantar fasciitis. Rest and corticosteroid injections may relieve the pain, but healing will result only from the correction of the biomechanical imbalance. When treating a hollow foot, it may be useful to insert a 6 mm thick, horseshoe-shaped felt pad or a wedge-shaped support under the patient's heel. On the other hand, when treating a prone foot, a flexible and laminated orthopedic appliance is recommended. In difficult cases, surgery may be required to remove the possible bony spurs.

Muscular sprains

Muscular sprains require 12–30 days of complete rest, together with the use of myorelaxing drugs and non-steroidal anti-inflammatory drugs (NSAIDs). Exercise can then be gradually resumed. Complete or incomplete rupture of the Achilles tendon is usually treated surgically, and it is advisable to operate as soon as possible. Incomplete rupture of the Achilles tendon is uncommon (**141**).

Stress fractures

Stress fractures require total rest. It is often necessary to put the leg in a cast, to protect the foot from bearing weight. The cast must be left on until there is complete consolidation of the fracture, which can be very slow. Fracture of the proximal fifth metatarsal may require four to eight weeks to consolidate.

141

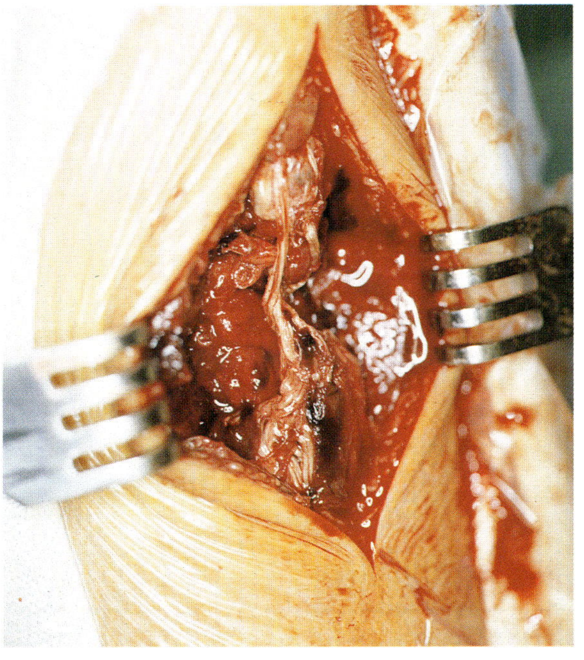

141 A ruptured Achilles tendon often appears frayed. *Courtesy of J. Vallotton.*

Conclusion

On the whole, the prevention and treatment of injuries to the foot in hiking or trekking is based on conservative measures. An appropriate choice of footwear is essential, as is the use of orthopedic appliances. Good physical condition is also necessary to prevent the injuries that have been described.

Muscular lesions

Alain Rostan and Charles Gobelet

Stiffness

When considering the range of muscular lesions, one often forgets that aching, stiff muscles are a form of injury. These symptoms appear a few hours after muscular effort has stopped (often on awakening the next day), and last for three to seven days. They often occur after a prolonged and rapid walk downhill, especially at the beginning of the season when the muscles are not yet accustomed to 'negative' working. This type of injury is characterized by a feeling of stiffness, by weakness of an algesic type and by pain on effort. The histological picture is particularly interesting: a clear injury situated in the contractile part of the myofibrillae is visible, especially at the level of the Z band (which appears blurred or is sometimes even totally obliterated) (**142**). This injury disappears with training of the muscles. It is important to remember that this is a true anatomical injury, even though it is purely microscopic.

 **142**

142 Electronic microscopy showing the biopsy of a striated muscle; enlarged Z striae with signs of rupture are visible.

Strained muscles

Strained muscles occur when the muscle is stretched beyond its physiological capabilities (not a true macroscopic, anatomical injury). The pain often appears progressively, and does not usually require the cessation of muscular effort. Factors which encourage strained muscles include muscular imbalances, unsuitable muscle training (possibly serious muscular stiffness, which has been allowed to develop), an inadequate warm-up, or the aftereffects of old injuries. Clinical examination shows localized pain or spasms; the muscular tension is reduced only slightly, and it functions normally in active movements, but the patient generally feels a little pain with contrary and with stretching movements. In treatment of the injury, one should first correct the muscular imbalances, and ensure that the muscle is allowed to develop under the best possible conditions, with sufficient hydration and, possibly, the use of magnesium. The use of physiotherapy is still controversial.

Microruptures

Moving up the scale of muscular injuries brings us to microruptures. The patient feels acute pain in the muscle during activity, although there is no associated 'snapping'; considerable pain at night ('hyperstiffness'), generally after uncommonly strenuous exercise; and pain that builds up progressively (over days or weeks), in relation to how often the patient repeats the particular type of exercise. With these injuries appear interstitial hemorrhagic suffusions, which are probably the result of microruptures of muscle fibers, revealed clearly by an ultrasonographic examination of the muscles (**143**). The second echographic criterion necessary for this diagnosis is the absence of true rupture of the muscles.

Two patterns of injury may be found: (i) intense, repeated physical effort, involving a muscle which has possibly been weakened by a previous injury or a retraction, producing microlesions; (ii) acute lesion of the muscle fibers, without true rupture (*see* 'torn muscles', p. 130). In the treatment of this type of injury, physiotherapy might reasonably be applied in the form of suction massages of the lymphatic drainage type, hydrotherapy (jet massage), electrotherapy with medium frequencies, short waves, ionizations (where applicable) and ultrasonics, and then progressive retraining of the muscles. Total recovery takes place within ten days to a fortnight.

143

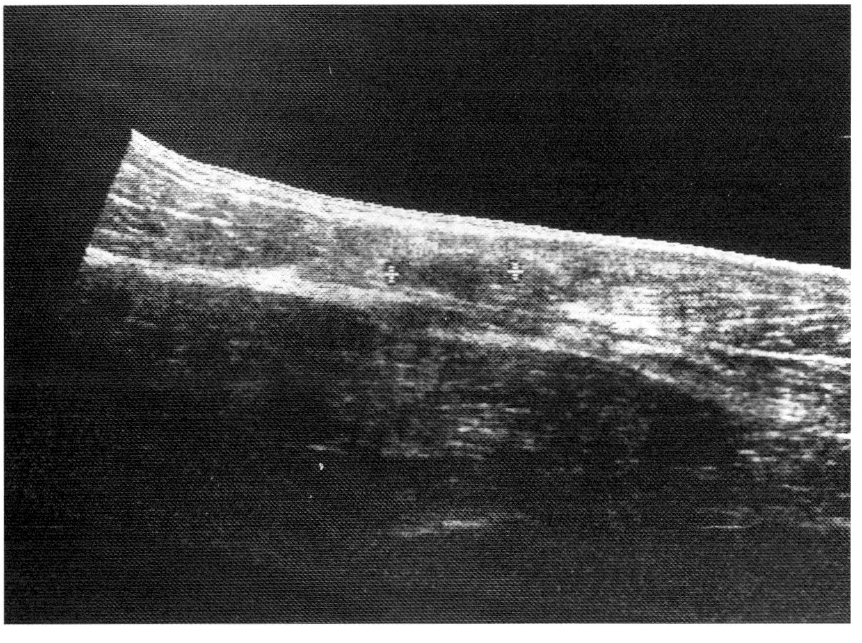

143 Echographic image of an acute muscular microrupture. The muscle is not ruptured, but the structure is disorganized (interstitial disorganization), with hyperechogenic lines at the site of the injury.

Torn muscles

A torn muscle results from the tearing of several muscle fibers, through an abrupt movement or direct impact. When the accident occurs, the patient generally hears a snapping sound. The pain is acute and rapidly becomes constant, forcing the patient to stop his activity.

Clinical examination frequently reveals a change in the shape of the muscle and an ecchymosis (absent in the case of microrupture), which appears 48–72 hours after the accident. Palpation causes sharp and often extensive pain. The muscular tension is reduced and active movements are painful and limited. An ultrasonographic examination (**144**) often enables the extent of the injury to be assessed. On the whole, the treatment described for microruptures is valid here, but each stage must be prolonged in view of the greater seriousness of the injury.

144

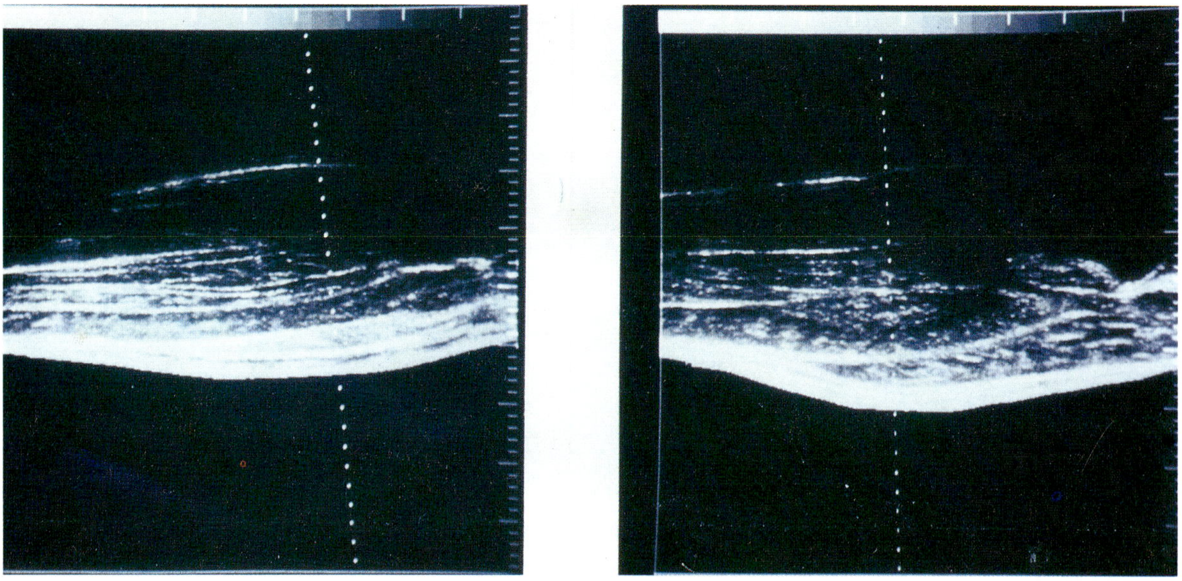

144 Partial muscular rupture in the medial part of the gastrocnemius muscle. Ultrasonography shows a small intramuscular rupture associated with significant interstitial edema. On the right is the opposite calf.

145 MRI of an interstitial hemorrhage in the left rectus femoris muscle.

145

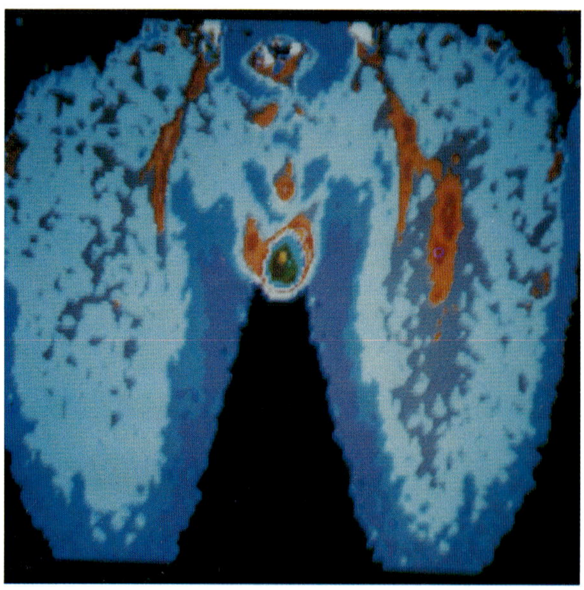

The first aim of this treatment is drainage of the muscle, to allow the muscle fibers to heal in the most favorable surroundings (elimination of local 'waste') (**145**). The period of recovery will vary from three to six weeks depending on the localization and the extent of the injury. When the calf is affected, one should always consider the possibility of a proximal detachment of the medial head of the gastrocnemius (clinical and ultrasonographic diagnosis), or of a rupture of the Achilles tendon, both of which may mimic torn muscles.

Ruptured muscles

These result from the rupture of several of the muscle's fasciculi, and sometimes of the entire muscle. A rupture is caused by direct impact on a contracted muscle, or by an acute, violent muscular effort. The pain appears sharply with a snapping noise, and is constant. Clinical examination reveals a 'hole' next to a globular swelling (retracted fasciculi). Palpation reveals a clear, very painful depression and sometimes local thickening of the muscle.

Muscular tension disappears and active movement is very painful and clearly greatly weakened. Opinions regarding the use of surgery and the point at which it becomes necessary to operate differ greatly from one region (and sometimes from one surgeon) to the next. The decision should be based on the extent of the lesion, its localization, and the age and activities of the injured person. Ultrasonographic examination gives a fairly accurate picture of the extent of the lesion, and often helps when deciding on the treatment (**146**).

146

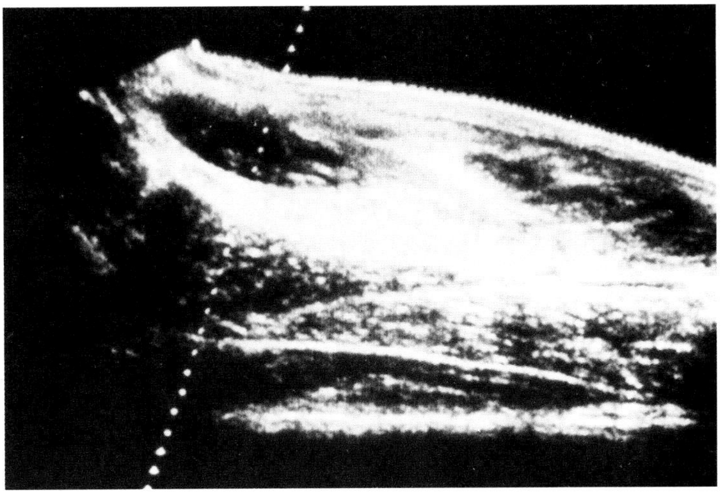

146 Echographic image of a partial but significant muscular rupture in the posterior part of the thigh, with considerable interstitial hemorrhage.

Muscular contusion

Contusion of a muscle can also vary greatly in severity from one case to another. It should be remembered that direct impact on a contracted muscle may cause dilaceration of the fasciculi fibers, with a significant interfascicular hemorrhage or even total rupture of the muscle. At the first examination, the possibility of a fracture must always be considered (for example, a fractured femur for contusion of the thigh muscles). After an accurate diagnosis has been made, the doctor should prescribe a suitable form of pain relief (and crutches if necessary), followed by physiotherapy, with a view to draining the area.

Intramuscular hematomas

147

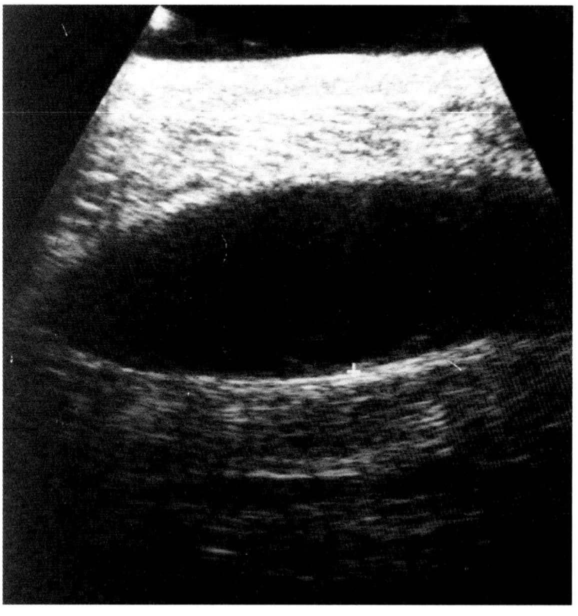

147 Characteristic echographic image of a significant localized hematoma of the left calf.

An intramuscular hematoma may occur when a patient is suffering from a microrupture (although this is rare), or from a torn, ruptured or bruised muscle. It is a local reaction, resulting in the accumulation of a liquid mass. Through ultrasonography, the extent of the hematomas may be assessed and, if necessary, they may be aspirated (**147**). This often speeds up recovery, but should not be undertaken lightly, because of the risk of secondary infection; the frequently preferred option of electrotherapy (short waves and ultrasonics) should also be discussed.

To summarize, muscular lesions may occur during any sporting activity, without any obvious connection with altitude, although hypoxia might possibly make the muscle cell more fragile (metabolic attack) and increase the risk of stiffness or microruptures occurring. When faced with an acute injury, the only recommended action is to cool the injury for a period of 20–25 minutes many times a day, in order to reduce the inflammatory reaction and local hemorrhage through vasoconstriction. A flexible bandage may be fitted where the injury is serious, again in order to restrict the amount of interstitial liquid. Pain relief should be prescribed, if necessary.

Ossifying myositis

Ossifying myositis is a possible secondary complication of muscular injuries. In its early stages, it is characterized by the presence of an inflammatory or pseudoinflammatory syndrome, with deep induration of the muscle which is not easy to delimit, sudden pains and even the sensation of local heat. At this stage, there is definite functional impairment. When the osteoma has formed, a hard mass can be felt extending deep into the muscle; this is generally not painful, but may restrict the function of the muscle because of its extent and localization.

Usually, there is no biological disturbance. Thermography reveals persistent hyperthermia, which generally regresses within three to six months. Echography shows a hyperechogenic image with a posterior cone of acoustic shade. Generally, radiography shows no injury at the initial stage, but from the fourth week, tissue calcification is visible (**148**).

Scintigraphy is the most suitable form of examination for early diagnosis, quickly revealing hyperfixation, associated with intense osteoblastic metabolism (**149**). Generally, scintigraphy shows a normalization within a few months. Magnetic resonance imaging also allows the extent of the injury to be assessed (**150**). However, this form of examination is used only exceptionally, because of its cost.

The differential diagnosis of ossifying myositis is circumscribed ossifying myositis, as Münchmeyer's progressive ossifying myositis is not relevant here. The cases of circumscribed ossifying myositis that we refer to are simple muscular calcifications, and also calcium accretions on the interstitial muscular connective tissue, in the absence of osteoid metaplasia (interstitial or circumscribed calcinosis: Thibierge–Weissenbach syndrome).

There are many factors that may be the cause of post-traumatic ossifying myositis, and all are still hypothetical: for example, an increased availability of calcium, or a genetic mutation or facilitation (HLA-B18).

Treatment is above all preventive: untimely mobilization and immobilization must be avoided. In the evolutive stage, anti-inflammatory radiotherapy may sometimes be applied. The preventive or curative administration of diphosphonates orally or intravenously is currently being studied, but results are still contradictory. Surgical curative treatment can be undertaken only in the late stage of the condition, when scintigraphic activity and the laboratory report are normal. Earlier surgery carries the risk of a relapse.

148 Ossifying myositis in the adductor muscles.

148

149 Scintigraphic image of ossifying myositis of the rectus femoris muscle, two months after a muscular contusion. This shows an important hyperactivity in the late phase.

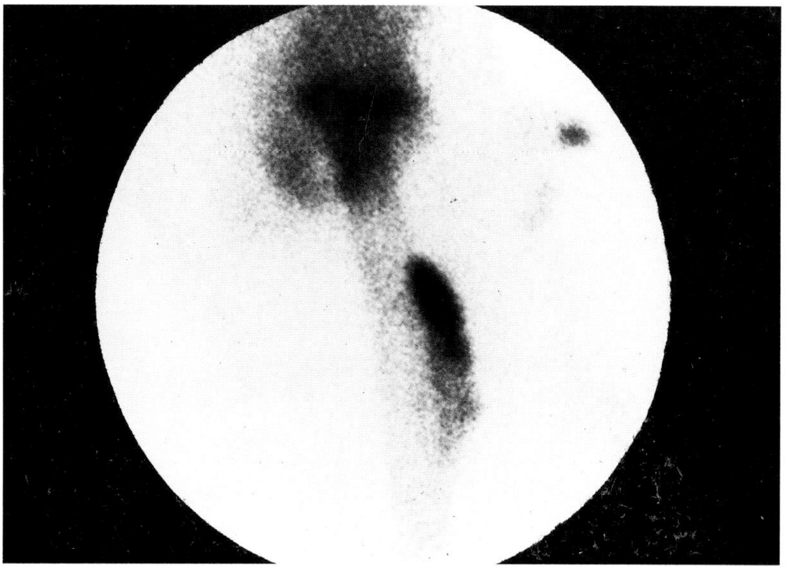

149

150 MRI of the case portrayed in **149**.

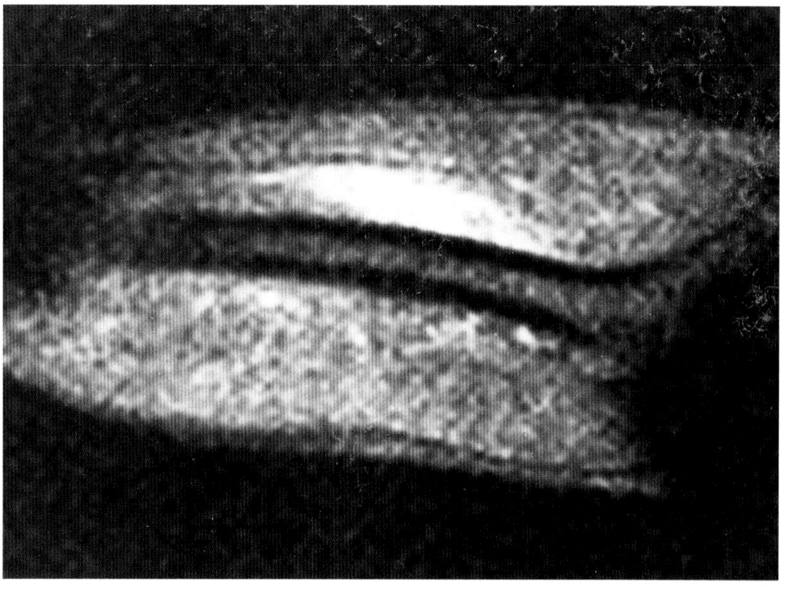

150

Downhill Skiing

Downhill skiing accidents: prevention and statistics

José M. Figueras, Aleix Vidal and Eulalia Escola

Skiing has evolved from what was once a means of transportation into one of today's most popular sports (**151**). It has undergone significant changes, and a relationship can be seen between these modifications and present-day skiing injuries. The changes are conditioned by a series of factors which may be placed in three main categories: the environment, skiing equipment and the skier's capabilities.

151

151 Crowded slopes are extremely dangerous and should be avoided.

The environment

The type of snow, the type of slope, the visibility, the weather conditions, the number of skiers and the general layout of the ski resort (signposting (**152**), length of queues, general information for skiers, etc.) may be determinant factors in accidents, and must therefore be taken into consideration if they are to be prevented.

The type of injury is often determined by the type of snow. Injuries to the lower limbs are more likely to occur in deep snow. Icy snow is the most common cause of accidents. However, on icy snow typical skiing accidents occur less often, with only 4% involving injuries to the lower limbs. When the conditions are icy, contusions and severe injuries can occur, especially on steep slopes where skiers cannot stop after falling. For this reason, this is considered to be the most dangerous type of snow by health care providers. It must be remembered that changes in the type of snow (icy snow turning into very heavy snow, for example) may occur within a few hours, especially towards the end of the season.

152

152 Good maintenance and clear signposting of runs are effective ways of preventing accidents.

Good slope maintenance in ski resorts improves safety and helps prevent accidents. However, it may also encourage the skier to go faster than his ability reasonably allows, and increases the risk of accidents due to collisions. Good slope layout and signposting are also vital, and the skiers must be kept informed of the condition of all runs, and of how to behave on them.

Poor visibility may somewhat increase the risk of injury. The sun and cold temperatures may themselves also cause injury, and skiers should be aware of the danger. It is safer to be too hot than too cold; it is better not to be suntanned than to be sunburned (even if less glamorous).

Skiing equipment

Changes in skiing equipment have altered some types of skiing injury and have, of course, helped prevent others. Some years ago, ankle injuries were quite common because of the use of short boots. As boots became higher legged, so did the site of the leg injuries (**153–157**). With short boots, the force which maintained the skier's weight on the inner edge of the downhill or outer ski (a basic principle in alpine skiing) resulted partially from considerable pronation by the lateral peroneal muscles. The high-legged boots have relieved the work of the lateral peroneal muscles and, consequently, dislocation of the lateral peroneal tendons has become less frequent; the work is now performed mainly by the adductor muscles of the thigh. However, high boots are the most significant factor in the increased incidence of knee injuries and transverse fractures of the tibia at the level of the top of the boot.

153

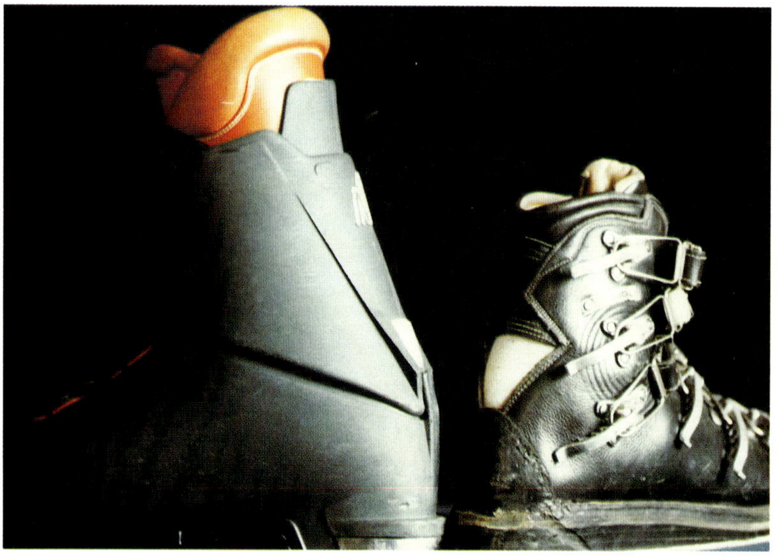

153 High- and short-legged boots.

154

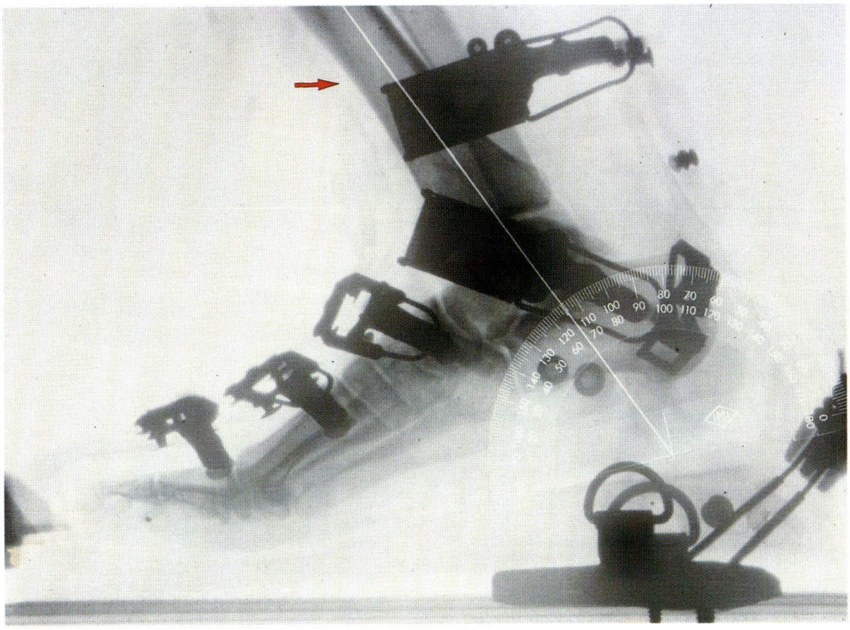

154 & 155 The modern high-legged boot limits the flexion of the ankle much more than the old short-legged one. Consequently, most leg injuries now occur at a higher level on the leg.

155

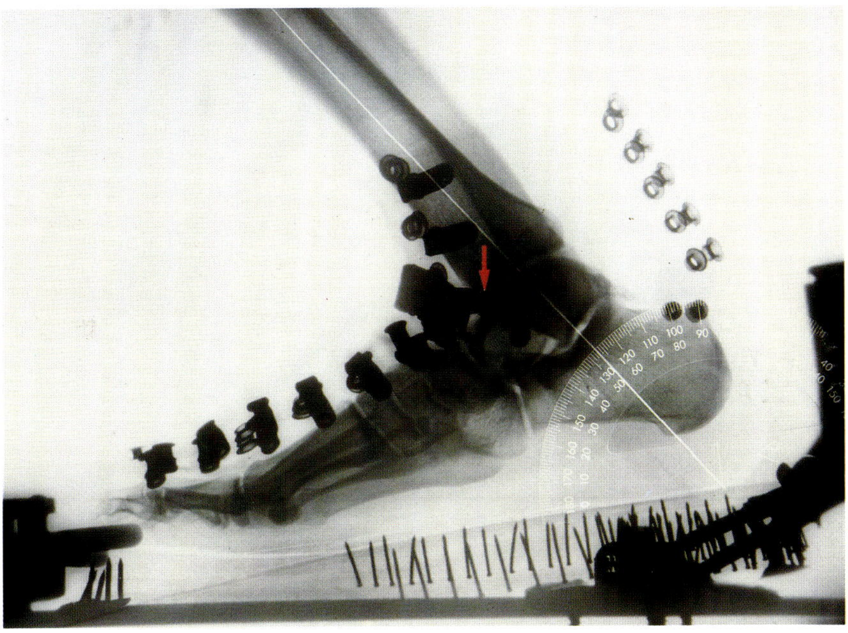

156

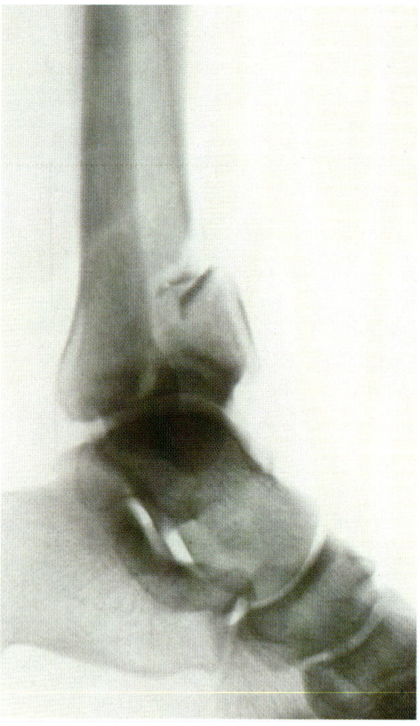

157

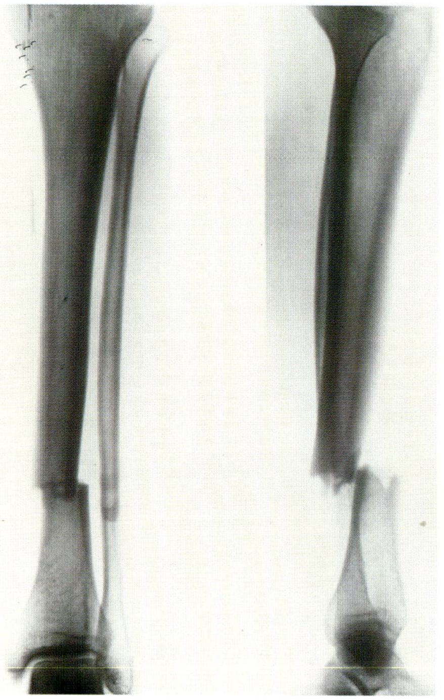

156 A characteristic intra-articular fracture of the distal tibia, due to compression against the talus in a short-legged boot.

157 Transverse leg fracture, due to compression against the top edge of a high-legged boot.

It is also interesting to note that higher-legged boots allow skiers, especially advanced ones, to recover from a backfalling position. In this dynamic situation, when the recovering force is important, an imbalance in the work of the thigh muscles (quadriceps and hamstrings) may occur, leaving the anterior cruciate ligament completely unprotected. This can cause its rupture, even without the skier ever falling (**158**).

Safety bindings (**159** & **160**) have become the most important means of preventing injury to the lower limbs. All the factors involved in correctly setting the safety bindings are important (age, sex, weight, skier's ability and size of the tibia's superior metaphysis). Every skier ought to be constantly concerned about the correct functioning of his safety bindings, and should check this by trying to release them himself. Many resorts (and equipment manufacturers) recommend an annual check/adjustment by a trained technician.

The skis are, of course, also extremely important. To allow them to slide in the safest and most satisfying way, their edges and bases should be in good condition. Their length must be chosen according to the weight and ability of the skier. For other skiers' safety it is also important that the ski stoppers function perfectly.

158

158 Nowadays, with the high-legged boot, the force is generated by the thigh and transmitted through the knee.

Proper eye protection (sunglasses or goggles) and hats should also be worn. Young children should always wear helmets. Hats are particularly important for skiers with long hair, to prevent it from getting caught in the ski and chair lifts. For the same reason, jackets and parkas should be properly buttoned or zipped.

159

160

159 Safety bindings are the most important factor in the prevention of ski injuries.

160 'Tryski': this device, developed by C. Humbert, joins the tips of the skis together and prevents them from either crossing or drifting apart. 'Tryski' can help prevent sprains in very young skiers and in beginners.

Skiers' capabilities

Injuries are frequently the skier's own fault. Physical fitness is of paramount importance on the slopes, yet many skiers do not take part in any other sports and lead a rather sedentary life between skiing trips. These skiers should take some other form of exercise to get into shape before skiing.

The skier should recognize when he is too tired to continue skiing, and should know when to stop. He should also be aware that the third day's skiing may be dangerous for physically unfit skiers, because of unconscious muscle fatigue. Each skier must ski according to his own level of ability. It is dangerous to ski alone and much better to ski in groups; the ideal is a group of about four skiers. When skiing in a group, racing is not advisable. When stopping, it is best to stand downhill from other, stationary skiers, and always on the edge of the ski run.

Regarding eating habits, it is best to have a high-calorie breakfast, a light lunch and a protein-rich dinner in the evening. It is important to note in relation to sex that women are more prone to ligamentary injury than men, as they tend to be less well protected by their muscles. Children pay poor attention to checking their safety bindings, and should be reminded to do so.

Statistics

Statistics on ski injuries vary from one ski resort to another. Some resorts have clinics, some have first aid facilities, and other still have to evacuate all their injured skiers. Statistics are often based on the number of injured skiers treated at first aid centers, and therefore do not include skiers who decide to get medical advice elsewhere, or those who think that the injury is unimportant. A typical example is thumb sprain, which is very common in skiers. Thus statistics available for some injuries may be inaccurately low. Furthermore, the authors believe that ski resorts ought to provide statistics that are related not only to the number of skiers but also to the amount of skiing done.

In **161** a comparison is shown of the incidence of alpine skiing injuries in La Molina (a ski resort in the Pyrenees) during two different winter seasons, ten years apart, in 1976–1977 and 1986–1987. The statistics show an increase in the percentage of injuries to the upper limbs and the head. Head injuries have increased from 9.7% in 1977 to 15.5% in 1987, and injuries to the upper limbs have risen from 20.6% in 1977 to 30.9% in 1987. On the other hand, injuries to the lower limbs have decreased from 66.6% in 1977 to 45.8% in 1987. These differences in the statistics are related to changes in the environment, in the skier's characteristics and in skiing equipment.

The authors have compared the statistics from 1985 in La Molina with those from the same year in Avoriaz, in the French Alps; the evolution of injuries is similar. Finally, the authors concur with Dr Marc-Hervé Binet (Avoriaz, France) that the main characteristic of ski injuries today is the increase in their severity.

161 Modification in the type of injury over a 10-year period.

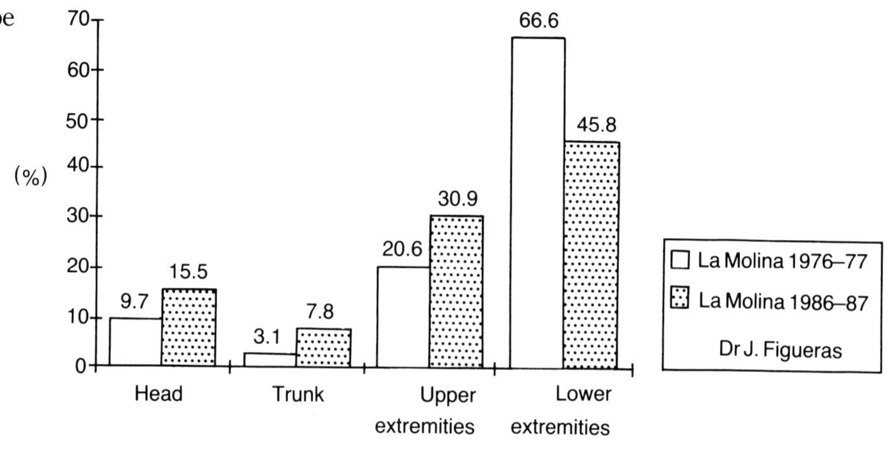

161

The management of skiing accidents

Henri Bèzes, Pierre Massart and Catherine Guyot

The city of Grenoble is located in the heart of the French Alps, in the immediate vicinity of many large winter sports resorts. In 1968, the city organized the Winter Olympic Games, for which a new University Hospital was built, the Hôpital-Sud (South Hospital). One of the authors, Henri Bezes, has been in charge of the Orthopedics and Traumatology Department of this hospital since 1968, and has tried to make it into a surgical center which specializes in skiing accidents.

Between 1968 and 1988, 17,120 people involved in skiing accidents have received treatment in the Emergency and Traumatology Department of the Hôpital-Sud; it is on these cases that the information contained in this section is based. In the 1968–1969 season, 250 skiers were treated, but the number has grown continuously and, since 1974, an average of 1,000 skiers has been treated each year. Most of the accidents occur as a result of downhill skiing, but over the last 10 years or so, cross-country skiing has become increasingly popular and is now responsible for 11–12% of all skiing accidents. More recently, monoskiing and snow surfing have further altered the statistics.

At the Hôpital-Sud, victims of skiing accidents are viewed as emergency patients and, wherever possible, receive definitive treatment from the start. This policy has necessitated certain changes in the staff's working hours, to enable the hospital to cope with the influx of accident victims throughout the afternoon. The whole medical team is therefore available to provide emergency treatment from 13.00 until late in the evening, including weekends and public holidays. To provide the best possible treatment and to make the subsequent transfer of patients to their home towns easier, osteosynthesis is often employed in the treatment of fractures in adults, using the A.O.–A.S.I.F. (Association Swiss Internal Fixation) equipment. The authors are convinced that this type of surgery has transformed the outcome of skiers' fractures by reducing their consequences to a minimum, both as far as the injury is concerned and, in particular, from a social and a professional point of view.

An analysis of the 17,120 cases treated in the Emergency and Traumatology Department of the Hôpital-Sud has allowed the authors to catalog the most common injuries that skiers sustain. The most significant changes in skiing accidents over the last 20 years or so have been, on the one hand, an increase in the number of injuries to the upper limbs, especially injuries to the shoulders and hands, and, on the other hand, an increase in sprained knees and a relative reduction in fractures of the legs.

The age of those injured in accidents ranges from two years to 79 (**162**). Just over half of the victims are referred to the hospital immediately by a resort doctor, while the others go there directly.

162 Age: incidence of ski injuries in the population.

162

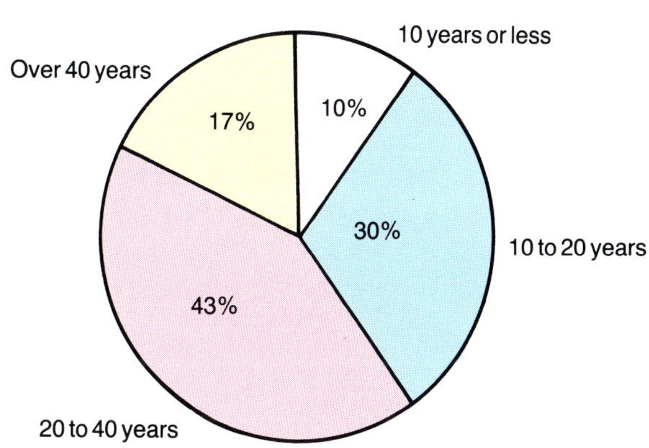

Of the 17,120 accidents discussed in this section, 206 should be considered separately; these are due to frostbite, wounds from slalom posts or barriers, and wounds from the edges of skis.

Frostbite. This injury is found only rarely among skiers in the Alps around Grenoble, and only 19 cases have been treated at the Hôpital-Sud over the last 20 years. The majority of these involve frostbite of the hands (12 cases) and are bilateral, with injuries to the fingers which are often symmetrical. The frostbite involves, in order of frequency, the fourth finger, the middle finger, the index and the little finger and, more rarely, the thumb.

Wounds from slalom posts or barriers. These injuries are sometimes very serious (**163**).

Wounds from the edges of skis. These generally result from collisions between skiers; they are clean cuts, which resemble deep knife wounds (180 cases) (**164**).

Of the remaining 15,497 injuries, there were 930 dislocations, 5,273 sprains and 9,294 fractures. Open fractures are very rare, and only 189 have been recorded (2%). In skiing accidents, the most common open fracture is a leg fracture with a generally modest opening situated on the anteromedial side (**165**).

163

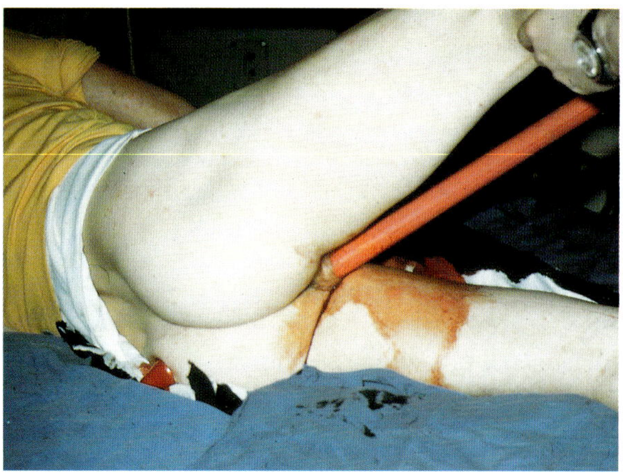

163 Injury due to impalement on a slalom stick, fortunately without perforation of the rectum.

164

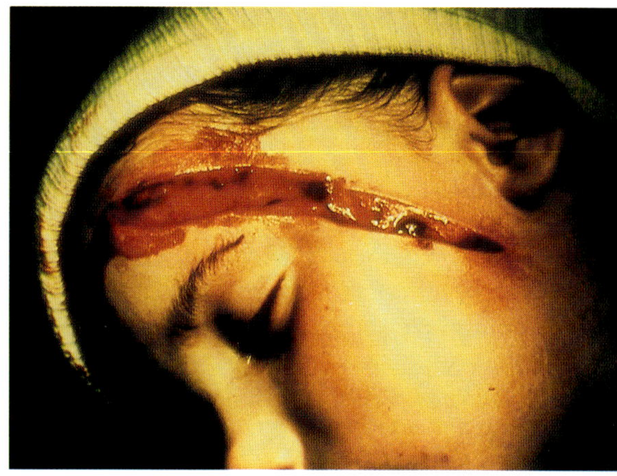

164 Facial wound caused by the edge of a ski.

165

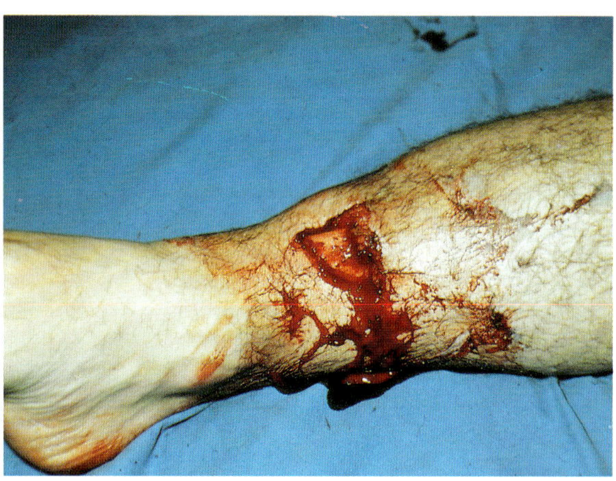

165 Typical open fracture of the leg: anterior cutaneous perforation by the proximal fragment.

As the Emergency and Traumatology Department of the Hôpital-Sud focuses on traumatology of the limbs, the skiing accidents treated there rarely involve the skull, the face, the spine, the thorax or the pelvis (1,668 cases). These injuries (discussed below) represent slightly less than 10% of the accident injuries as a whole.

Trauma of the skull and face. This generally results from a collision between two skiers, or a blow from the seat of a chair lift or from the pole of a drag lift. Fractures of the nose are common.

Spinal trauma. This often involves several vertebrae. The fractures (216 cases) are rarely located at the cervical level, being more common at the lumbar level, although most occur at the dorsal level or at the dorsolumbar junction.

Thoracic or thoracoabdominal trauma. This generally involves fractures of the ribs or bruising of the kidneys.

Pelvic trauma. At the Hôpital-Sud, the authors have surgically treated 10 fractures of the pubis and 15 fractures of the acetabulum.

When necessary, the patient is taken either immediately or secondarily to the specialist departments (neurosurgery, maxillofacial surgery, urology) of the Hôpital-Nord (North Hospital).

Injuries to the upper limbs (4,231 cases)

More than one skier in four who is involved in a skiing accident suffers from an injury to the upper limbs. The injuries are located mainly in the scapular area and shoulder (1,782 cases), and in the wrist and the hand (1,827 cases). Other injuries occur to the elbow (277 cases) and the forearm (195 cases). Several fractures of the humeral diaphysis (124 cases) have also been recorded. The percentage of injuries to the upper limbs has increased progressively during the past 13 years, rising from 12% of the overall number of skiing injuries in 1974, to 24% in 1987. This can be explained by the improvement in skiing equipment (safety bindings in particular), which has resulted in injuries at higher speeds and stronger impact, which mainly affect the upper limbs. To this can be added the increase in cross-country skiing, which has influenced this change in percentage, as has the relative reduction in fractures of the leg.

Injuries to the shoulder and the scapular area

These injuries alone (1,782 cases) represent 42% of the injuries to the upper limbs.

Scapulohumeral dislocation. This is a common injury (554 cases) (**166 & 167**), and may be either simple, or complicated by fracture with dislocation of the trochiter (156 cases). The dislocations are nearly always anteromedial, and rarely erect (11 cases) or posterior (11 cases). Posterior dislocations are sometimes difficult to recognize and care must be taken in the diagnosis. Of the total number of dislocations, 12% are recurrent. No vascular complications were observed by the author, though there were nine cases with neurological complications, with circumflex paresis or paralysis and/or fairly serious stretching of the brachial plexus.

166 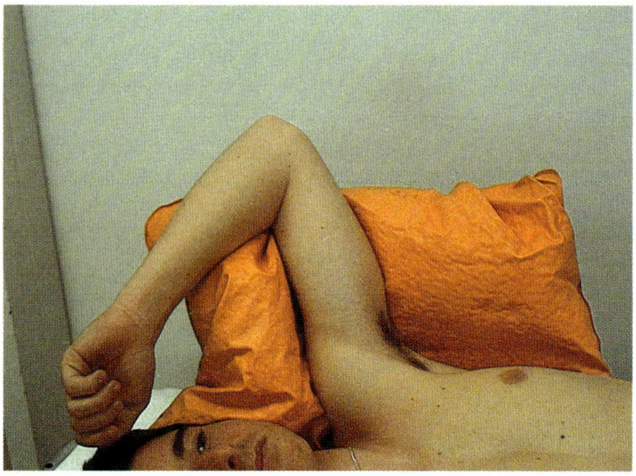 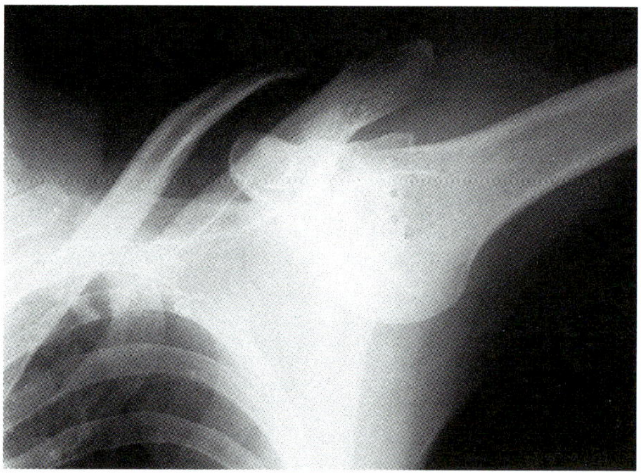 167

166 & 167 'Erecta' scapulohumeral dislocation of the left shoulder.

Fractures of the upper end of the humerus. These fractures (475 cases) are best treated with a hanging cast, or by immobilizing the elbow close to the body. Fifty fractures were operated on, mainly fracture–dislocations in which the head of the humerus, which was separated from the diaphysis, could not be reinserted into the glenoid cavity using external maneuvers; hence the need for open surgical reinsertion and the setting of the fracture using a plate.

Fractures of the clavicle. Several of these fractures, of which there are 373 cases, were operated on, whether transverse or short-oblique fractures, or complex fractures with one or two intermediary fragments.

Acromioclavicular dislocations. These (133 cases) are operated on only where there is considerable diastasis (39 cases) (**168**).

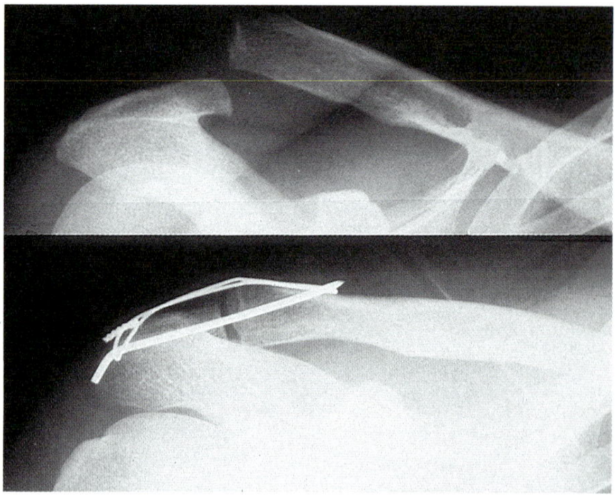 168

168 Grade III right acromioclavicular dislocation before and after surgical repair.

Injuries to the wrist and hand

These injuries (1,827 cases) represent over 43% of the total number of injuries to the upper limbs. Many result from cross-country skiing accidents. A third (633 cases) are wrist injuries.

Radiocarpal dislocations or **isolated fractures of the bones of the carpus.** These are rare; **epiphyseal detachment** of the lower end of the radius in children is more common, and is generally accompanied by considerable displacement.

Fractures of the distal end of the radius. These fractures (428 cases) are 3.5 times more common in cross-country skiing accidents than in downhill skiing accidents. Intra-focal pinning using the Kapandji method is often employed in the treatment of Colles' fractures (251 cases)

(**169**), and osteosynthesis using plates is almost systematically used in the treatment of Smith's fractures (29 cases); these techniques prevent secondary displacement, which is almost inevitable when the limb is immobilized in plaster.

Injuries to the hand (1,194 cases) are increasing constantly.

Sprains of the metacarpophalangeal joint of the thumb. These sprains (662 cases) mainly involve the cubital collateral ligament. They generally occur when the strap of the ski pole, which passes round the joint of the thumb, is pulled sharply, or when the skier falls and lands awkwardly with his thumb stuck in the snow. A thorough clinical and radiological examination should be performed. It was necessary to operate on more than one-third of these sprains, to avoid a residual disabling instability (*see* p. 155).

Fractures of the metacarpals (268 cases) or **fractures of the phalanges** (98 cases). These are equally common in the column of the thumb and the four other radii. A large number of these fractures required surgical treatment (**170**).

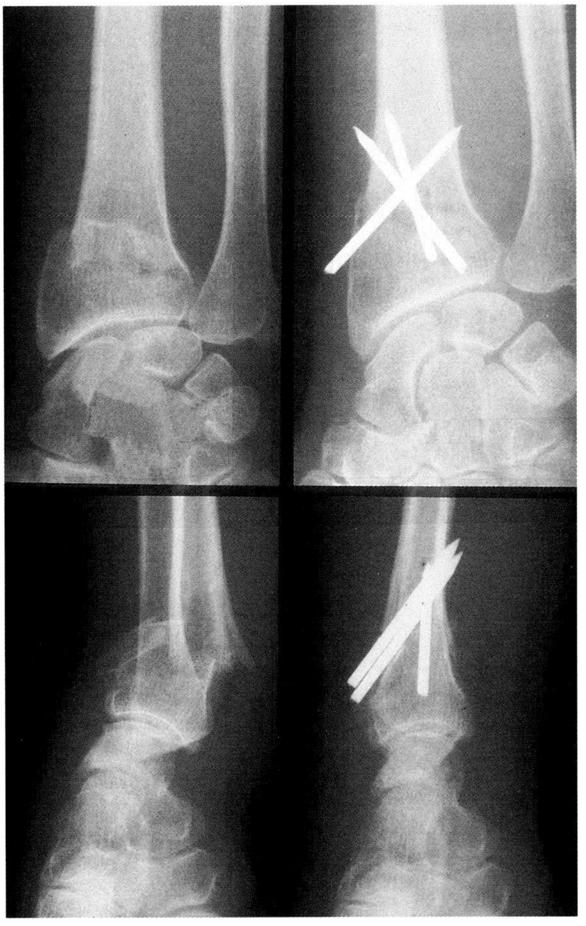

169 Kapandji management of Colles' fracture.

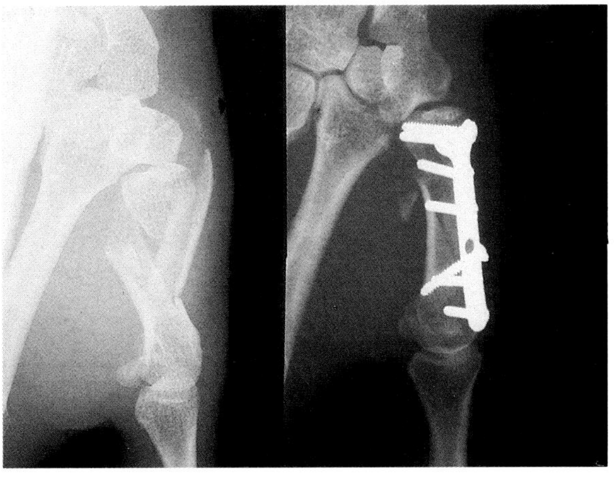

170 Comminuted fracture of the right first metacarpal bone, treated using the A.O.–A.S.I.F. osteosynthesis technique.

Injuries to the elbow and to the forearm

Injuries to the elbow. Of the 277 cases of elbow injury, 79 are dislocations, all posterior and with an occasional associated fracture of the epitrochlea, and 128 are fractures, which were mostly treated surgically.

Injuries to the forearm. These injuries (195 cases) are partly low metaphyseal fractures, which occur mainly in children and are treated orthopedically, and partly diaphyseal fractures, which usually require surgery.

Fractures of the humeral diaphysis

Over half of these fractures (124 cases in total) were treated surgically (**171** & **172**). It is important to use a technique with avoids contact with the radial nerve, thereby minimizing the risk of postoperative paralysis. Anatomical reconstruction is obtained by means of a stable, solid setting of the bone; this avoids the need for immobilization and consequently has the advantage of an early recovery of all the limb's functions, unlike orthopedic treatment using a thoraco-brachial plaster cast or a hanging cast.

171

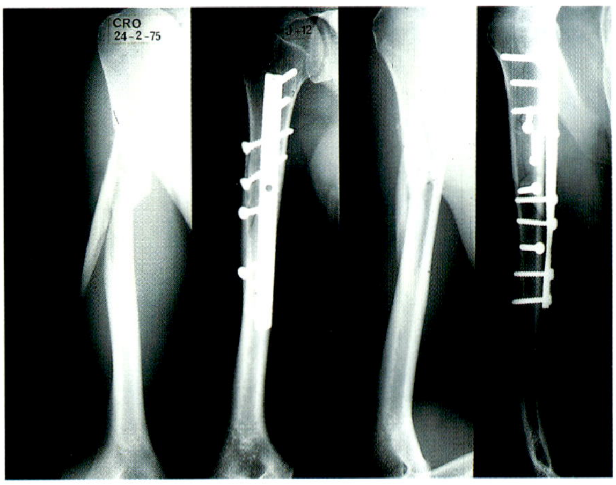

172

171 & 172 Comminuted humeral shaft fracture, treated using the A.O.–A.S.I.F. osteosynthesis technique. Function was regained 13 days later.

Injuries to the lower limbs (11,584 cases)

As one might expect, injuries to the lower limbs make up the vast majority of all skiing injuries, and constitute over two-thirds of the overall total (11,584 out of 17,120). Certain injuries affect only the soft parts of the lower limbs (mainly the tendons), and others are osteoarticular.

Injuries to the tendons

Dislocation of the peroneal tendons. There are 20 recorded cases of this.

Rupture of the Achilles tendon. These ruptures (58 cases in total) are generally repaired surgically (48 cases). The injury, which was common 20 years ago, has now become very rare with only three cases a year at the Hôpital-Sud. The rigidity and height of today's ski boots have much to do with this reduction, as they restrict the bending of the ankle.

Osteoarticular injuries

On account of the violence of certain skiing injuries, due to increasingly high speeds, osteoarticular injuries to the proximal part of the lower limb are becoming more and more common.

Injuries to the proximal part of the leg

Hip dislocation. There are 60 recorded cases (**173** & **174**).

Fractures of the upper end of the femur. Most of these fractures (158 cases in total) (**175** & **176**) require surgical treatment. Moreover, subsequent injuries to the lower limbs, caused by cross-country skiing, often occur at the site of these fractures.

Fractures of the femoral diaphysis. These injuries (374 cases) are six times more common in skiers than ruptures of the Achilles tendon. While the fractures are usually treated orthopedically in children, fractures in adults must be operated on, and the authors generally perform osteosynthesis using a plate.

Three types of injuries, which are located below the femur, greatly dominate the traumatic pathology of skiing, involving 10,793 cases out of a total of 17,120. These are: knee injuries (3,882 cases); ankle and foot injuries (1,811 cases); and broken legs (5,013 cases).

173

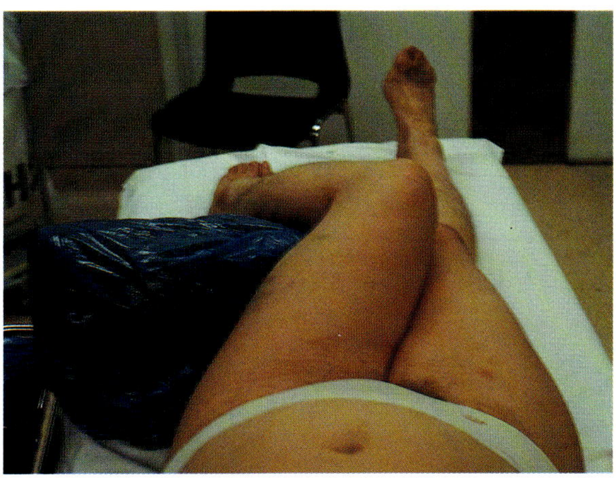

174

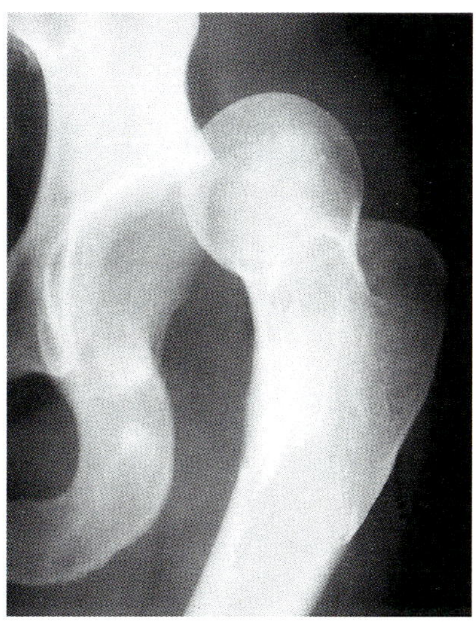

173 & 174 Dislocated hip in a female cross-country skier.

175 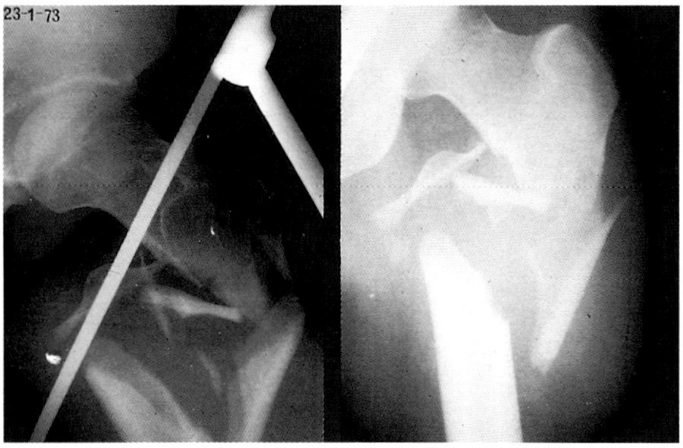

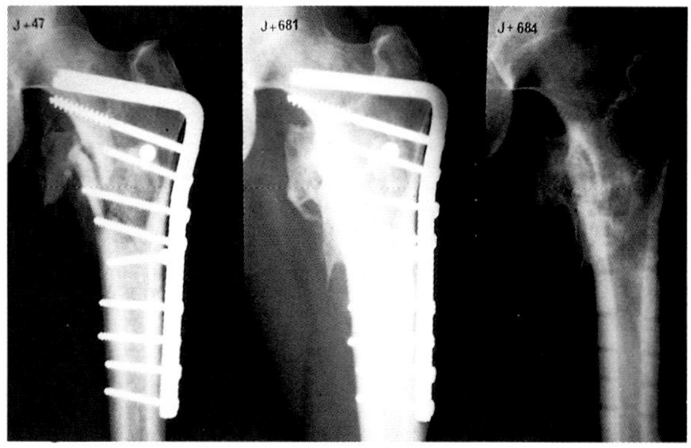 176

175 & 176 Complex and open comminuted fracture of the proximal left femur, treated using the A.O.–A.S.I.F. osteosynthesis technique and follow-up.

Knee injuries

Meniscal tears. These (43 cases) are isolated with hemarthrosis.

Dislocations of the knee. These (13 cases) are occasionally the cause of neurovascular injuries. Unlike dislocations of the hip, the elbow or the shoulder, dislocations of the knee, which may be reduced in emergencies by external manipulation, must subsequently be operated on to repair surgically and precisely the injuries to the central pivot (the anterior and posterior cruciate ligaments), to the meniscus and to the internal or external capsuloligamentary structures. Surgery should be carried out as soon as possible, as substantial damage can result from a dislocated knee.

Intra-articular fractures of the proximal tibia. In these fractures (116 cases) the bone is generally smashed in and split; they are more common on the lateral joint surface.

Dislocation of the patella. These dislocations (38 cases) are generally reduced by means of external manipulation, and any realignment of the extensor structures of the knee is performed at a later time.

Fractures of the patella. Of the 35 cases, 19 were treated surgically.

Sprained knees. The 3,480 cases of this injury constitute 20.3% of all the patients seen in our department for the 1968–1988 period. Sprained knees have become more and more common in recent years: from 1968 to 1974, there were less than 100 per season; between 1975 and 1979, there were between 100 and 200 per season; and from 1980 to 1988, between 200 and 300 per season, with the incidence reaching its peak at 319 cases in the 1983–1984 season (**177**).

1,136 minor sprains were treated with ordinary bandages, elastic bandages or removable splints. 1,693 sprains of moderate seriousness were treated with a cruromalleolar plaster cast, which was kept on for five to six weeks. Of 651 serious sprains, with definite clinical or radiological signs of injury to the ligaments, 460 required surgical treatment. In fact, for the past 15 years, the department has followed the principle of treating such ruptures of the ligaments as emergency or semi-emergency situations, and repairing them immediately. Amongst these serious sprains are 368 injuries to the anterior cruciate ligament, 327 injuries to the internal capsuloligamentary structures, and 47 injuries affecting the posterior cruciate ligament or the external capsuloligamentary structures, or appearing as fractures in the tibial spines. These statistics show that, in knee sprains in skiers, it is the anterior cruciate ligament and the internal lateral ligament that are most frequently injured, either alone or jointly (*see* p. 152).

177

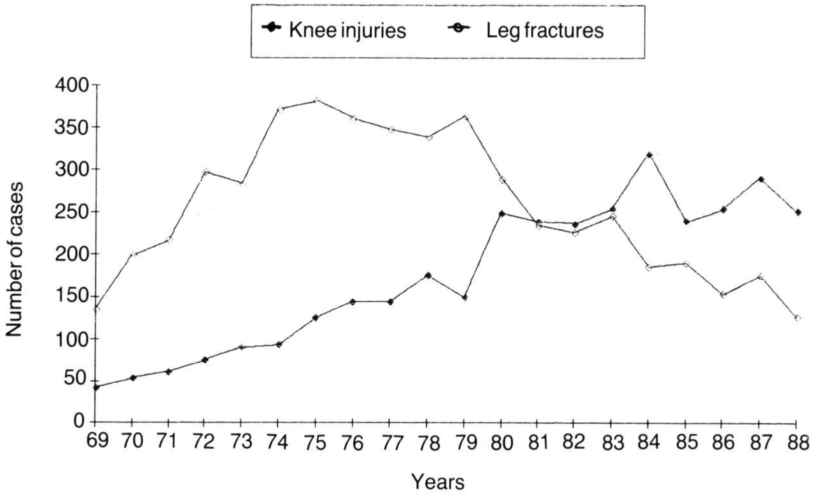

177 Evolution of the number of knee injuries and fractures of the leg.

Ankle and foot injuries

178

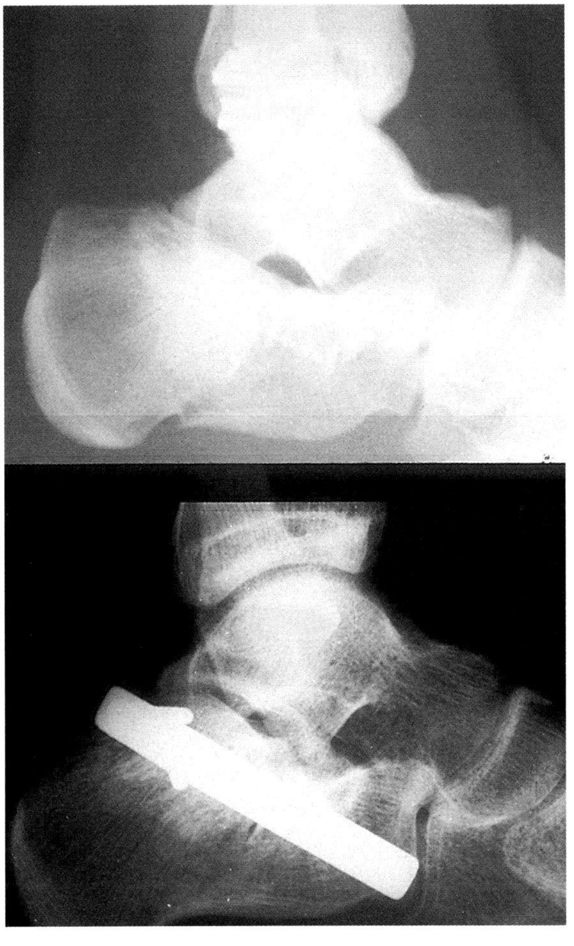

Injuries to the feet are extremely rare due to the good protection provided by ski boots. Thus, the department's statistics show only nine **fractures of the talus** and 22 **fractures of the calcaneum**; these fractures are usually operated on using a technique developed in our department, to minimize thalamic depression (**178**).

Injuries to the ankle are far more common, and are listed below.

Mild sprains of the external lateral ligament. There are 742 recorded cases of this injury.

Fractures–dislocations. The 973 recorded cases of this type of injury include: one pure tibiotarsal dislocation; 41 epiphyseal separations of the lower end of the tibia or of the external malleolus; 160 fractures of the tibial shaft; 768 fractures of the malleolus – these are generally operated on

178 Calcaneal fracture, treated using Bezes' original technique with a ⅓ tube plate on the lateral side.

immediately if they are displaced, with both the injuries to the bones and the injuries to the ligaments being repaired simultaneously. The most common skiing injury of this type is a short chamfer fracture of the external malleolus, with tearing of the internal lateral ligament (**179**).

179 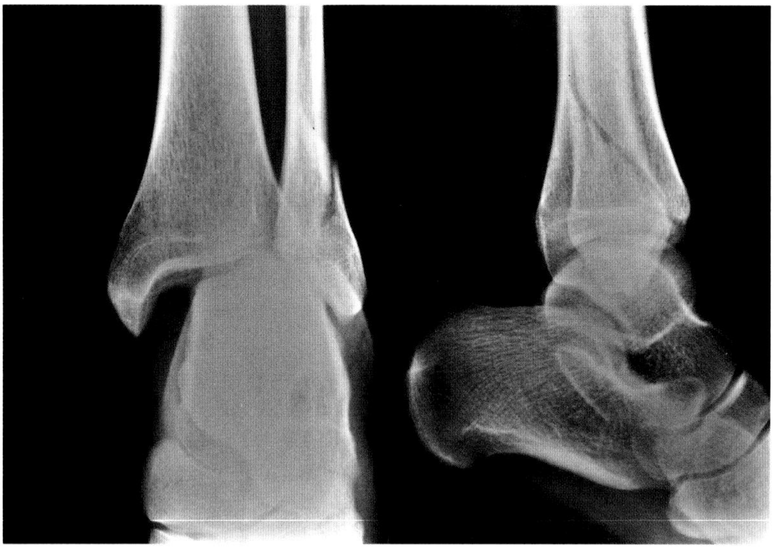

179 Typical valgus ankle fracture in cross-country skiing: rupture of the medial ligament and external malleolar fracture at the level of the peroneotibial syndesmosis.

Broken legs

Broken legs are the most common injury resulting from skiing accidents, making up 30% of the total number of such injuries. Until 1978, they represented 40–45% of the total number of injuries treated in our department, but their number has progressively decreased and, at present, they represent less than 20%. The increasing number of sprained knees has greatly influenced this change in percentage, as have improvements in skiing equipment, particularly ski boots and safety bindings.

Out of 5,013 fractures, 55 were bilateral and 147 were open (**180**). This 'avalanche' which comes crashing down onto our department every season raises certain problems, particularly when, as often happens, the victims come from outside the region. Therefore, it has been necessary to set up a policy for the treatment of broken legs.

The policy followed with children is to reduce the fractures by means of external manipulation, and then to send the patients back to their home towns, with their legs immobilized in plaster; only those failing to respond to orthopedic reduction are then treated surgically; the 2,170 cases of broken legs in children aged under 15 were nearly all reduced orthopedically, generally using a Boppe frame (with help from a transcalcaneal traction pin and a proximal extension pin); they were then immobilized in a cruropedal plaster cast, with the pins sunk into it (**181**).

If the same policy had been followed with the 2,843 adults with fractures, this would, of course, have entailed the risk of a large number of these fractures being replastered or operated on a few days or a few weeks later, either because of a possible secondary displacement, or perhaps due to the enterprising nature of the surgeon consulted. In order to avoid these setbacks, which are often unpleasant for both the injured person and the first surgeon, who is held responsible for the poor, or allegedly poor initial setting of the injury, the authors often opt for surgery. Thus, out of a total of 5,013 fractures, 2,716 were treated orthopedically (including 2,103 cases involving patients under 15 years of age), while 2,297 were treated surgically (only 67 of these were children aged under 15); 83 cases were treated using nailing, 68 with simple screwing, and 2,146 by means of screwed plates (**182**). This latter figure is a strong indication of the trust our department places in this

180

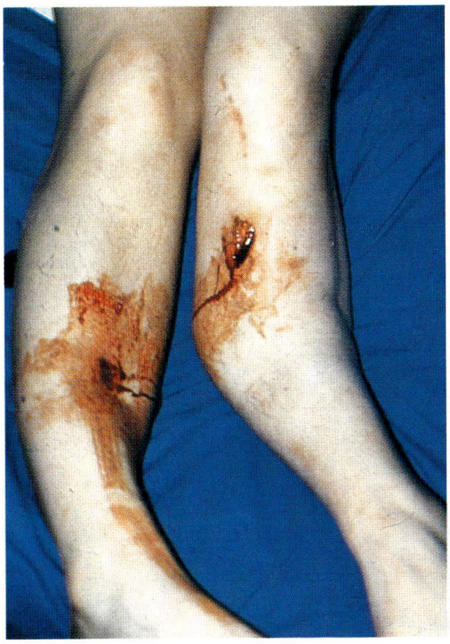

181

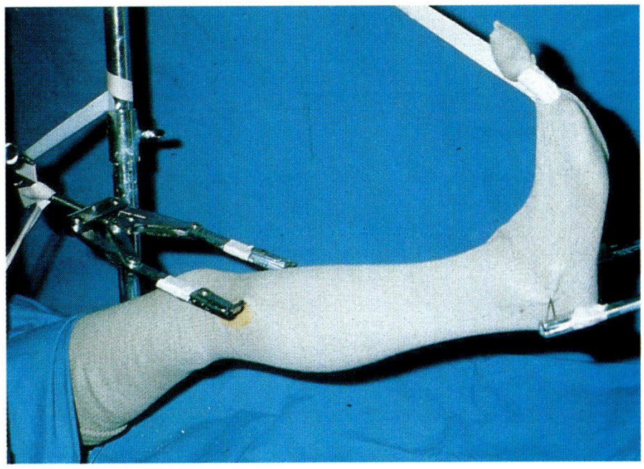

180 Open fracture of both legs.

181 Orthopedic treatment of a leg fracture in a child.

method, which has proven ideal for treating leg fractures sustained while skiing. While there have been some major complications (infection, pseudoarthrosis or refracture after ablation of the plates), these have been so few in number as to be negligible in comparison with the enormous benefits and remarkable results produced by this method. To avoid local surgical trauma, these forms of osteosynthesis must be performed with great care as, contrary to appearances, they involve delicate and often difficult surgery.

The main advantage of osteosynthesis is that no immobilization is necessary (**183**). The leg is kept in a raised position on a foam splint for a few days, after which the patient quickly regains control of the limb, and is able to make his own way home. Generally, the patient is able to drive a car six weeks after the operation, and stop using his crutches after two or three months.

182

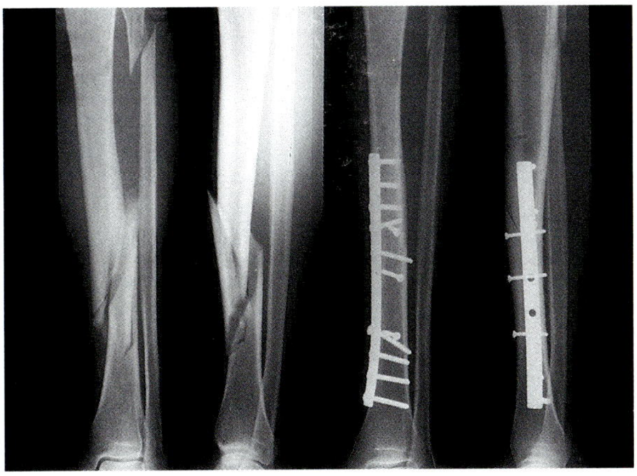

183

182 Comminuted fracture of the leg, with A.O.–A.S.I.F. osteosynthesis of the tibia.

183 Osteosynthesis provides a rapid mobilization and functional postoperative treatment of fractures.

Injuries to the ligaments of the knee

Pierre-François Leyvraz

The knee joint is one of the most vulnerable to sprains (**184**). As a hinge joint between two long lever arms, it can be subjected to excessive strains in some circumstances. In standing, the forces on the knee joint are mainly compressive, acting along the vertical axis of the lower limb, coapting the joint and thereby increasing its stability. During movement, on the other hand, horizontal shearing forces and movements are produced, tending to separate the tibial and femoral surfaces. This natural tendency to instability is favored by the anatomy of knee joint, in which the convex form of the femoral condyles is placed in contact with the flat articular surfaces of the tibia to produce a highly incongruent articulation.

The unusual anatomy of the knee joint and the strain under which it is placed necessitate strong complementary stabilizing mechanisms, able to control the joint's natural tendency to instability. There are two such mechanisms: the **active muscular system and the passive capsulo-meniscal and ligamentous system.** The first is by far the most powerful and, in normal circumstances, the most commonly used.

The balanced contractions of the various groups of muscles around the knee ensure constant coaptation and optimal stability of the joint. The distribution and intensity of the muscular contractions depend on proprioceptive receptors within the ligamentous system. These receptors are sensitive to stretching of the ligaments, and trigger off the appropriate muscular response when stimulated. The reflex activity of the muscles immediately relieves, and therefore protects, the stretched ligament. However, this regulatory mechanism of self-protection is very fragile and not without its limitations, namely in the speed of its response. Very fast movements and strong external forces, as are found in some sporting activities, can take the mechanism by surprise, not allowing it enough time to react and to carry out its protective role. This is even more likely to occur if the reflex is not triggered off or if the

184

184 The knee is the most vulnerable joint in downhill skiing. *Courtesy of B. Brand.*

musculature is underdeveloped. In these circumstances, the ligamentous system becomes isolated and must single-handedly maintain the joint's stability. If the external forces acting at the moment of impact are weaker than the ligament's resistance, the stability is maintained; but if they are stronger, the ligament tears, resulting in a **sprain**.

The severity of the sprain is determined through examination of: (i) the histological lesions, by analyzing the degree of structural damage to the torn ligaments; (ii) the macroscopic, anatomical lesions, by recording the number and the type of torn ligaments.

Each ligament is made up of several parallel collagen fibers, linked by tenuous interfibrillary bridges. In a **first degree sprain**, all the fibers are stretched to their limit, but are not ruptured. Some interfibrillary bridges are broken, but the ligament as a whole is not altered and its mechanical characteristics remain intact: after relaxation, it returns to its original length. In a **second**

degree sprain, together with several broken interfibrillary bridges, there are also some torn longitudinal fibers in the body of the ligament, which cause it to loose its normal elasticity: the ligament remains partially stretched with a residual slackness, discernible through clinical examination. In a **third degree sprain**, all the fibers rupture, leaving a macroscopic discontinuity in the ligament. The ligament is unable to function and there is considerable residual slackness. The degree of severity of the sprain is directly proportional to the intensity of the external force applied to the ligament.

Knee ligaments can be separated into two rough anatomical groups (**185**): **peripheral ligaments**: medial and lateral collateral; and **ligaments of the central pivot**: anterior and posterior cruciate. The collateral ligaments receive a good blood supply. Therefore, they heal well when correctly immobilized. However, the cruciate ligaments are stretched across the center of the joint cavity, and are immersed in synovial fluid. They never heal spontaneously after tearing, and their torn ends atrophy slowly if left unattended. Thus the severity of the injury depends on the healing capacity of the ligament. Sprains involving the ligaments of the central pivot are always serious; without surgical treatment, the ligament stays permanently slack. However, sprains involving the peripheral ligaments may heal without any permanent symptomatic slackness and thus are benign.

In a sprain, one or more ligaments can be injured to a varied histological degree. Any combination of injuries is possible, though some are more common than others. These depend on the intensity of the external forces applied to the ligament, on their direction, and on the position of the knee at the moment of impact. The most serious injury is total (third degree) rupture of all the ligaments, with dislocation of the knee. The most benign is a simple distension (first degree) of a single peripheral ligament. The most common sports injury is a complete rupture of the medial collateral ligament, together with tearing of the medial meniscus and a possible first or second degree lesion of the anterior cruciate ligament. This combination constitutes the injury formerly named by O'Donoghue as the 'unhappy triad' (**186**).

185

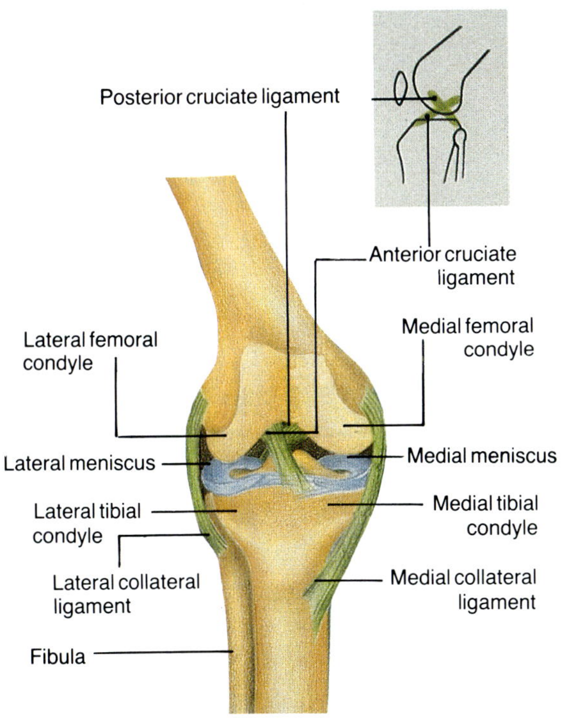

185 Anatomy of the knee.

186

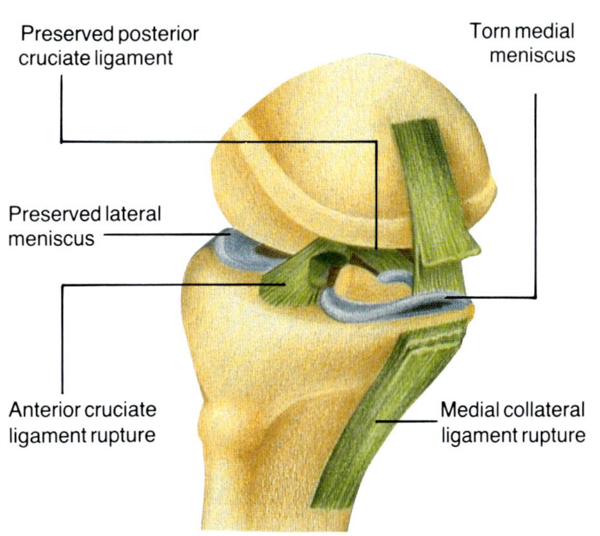

186 O'Donoghue's 'unhappy triad'.

Because of the great variety of injuries of this type, the examiner can rarely deduce the extent and severity of the sprain simply by knowing the type of sport involved or the circumstances of the accident. Only a thorough clinical examination, sometimes accompanied by other forms of investigation (arthroscopy, arthrography, standard X ray or magnetic resonance imaging), will provide a full picture of the injuries and allow a complete diagnosis to be made. A thorough diagnostic analysis is crucial for the selection of an appropriate form of treatment.

This selection is difficult and does not depend only on the anatomical lesions, but also on the patient's age, morphology, sporting abilities and ambitions. The choice is made harder by the existence of several treatment programmes for each type of injury, varying from country to country and from one surgeon to another. However, certain guidelines (listed below) can help the clinician to make his treatment choice.

- Two-thirds of untreated anterior cruciate tears lead, in subsequent months or years, to a symptomatic rotational instability, which tends to increase with time and to bring on secondary pathology (meniscal tears, osteoarthritis).
- Repairs of ligaments of the central pivot have a good chance of healing when the tear is at the bony insertions; they are more problematic if the central part of the ligament is affected (**187**).
- Isolated ruptures of the posterior cruciate ligament, unlike those of the anterior cruciate ligament, lead to secondary osteoarthritis only belatedly.
- With adequate immobilization, isolated peripheral ruptures heal well, and rarely result in long-term troublesome instability.
- Immobilization in plaster for more than six weeks causes muscular atrophy and cartilaginous damage, which can be permanent.
- It takes several months for a torn ligament to recover its normal mechanical characteristics.

When treating this type of injury, the physician should consider these observations and draw on his own personal experience to find the best, most suitable compromise for his patient, choosing from the various forms of treatment available (functional, plaster cast, surgery).

Neglecting a knee sprain can result in unfortunate consequences in the mid- or long-term; for example, time and time again, failure to diagnose

187

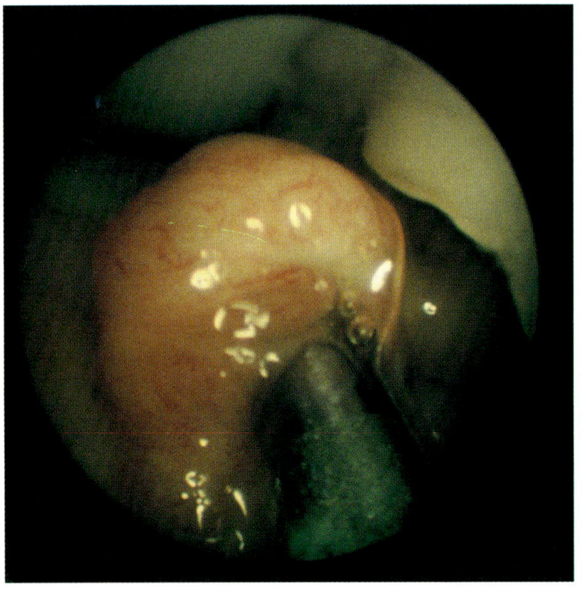

187 Rupture of the anterior cruciate ligament (arthroscopic view).

anterior cruciate tears results in rotational instability, preventing sporting acitivities and requiring a secondary ligamentary reconstruction. Even when perfectly carried out, full function of the knee is not always restored, and the result is often less satisfactory than that produced by a good primary repair. This type of situation is made all the more likely by the fact that sprain symptoms are often unreliable indicators of the real severity of the injury: a second degree medial collateral sprain (a benign lesion) is extremely painful, and the significant local hematoma that often appears with it can suggest a serious injury; on the other hand, an isolated anterior cruciate tear (a severe injury) can be asymptomatic. Thus, even when faced with an apparently mild knee sprain, a fast but thorough clinical examination is never superfluous, and is always better than holding back and thereby missing the opportunity of applying the correct form of treatment.

Capsuloligamentary injuries to the metacarpophalangeal joint of the thumb ('skier's thumb')

José Cantero and Jacques Vallotton

The rupture of the cubital capsuloligamentary apparatus of the thumb's metacarpophalangeal joint is one of the most common injuries in skiing (**188**). Neglect of a serious sprain results in the loss of an effective digitodigital grip between the thumb and the index finger. This handicap alone greatly reduces the function of the hand, and the results of secondary repair are often disappointing.

This section seeks to explain the mechanism of this injury, its radiological and clinical examination, and to define the treatment plan and the indications for surgery.

◀188

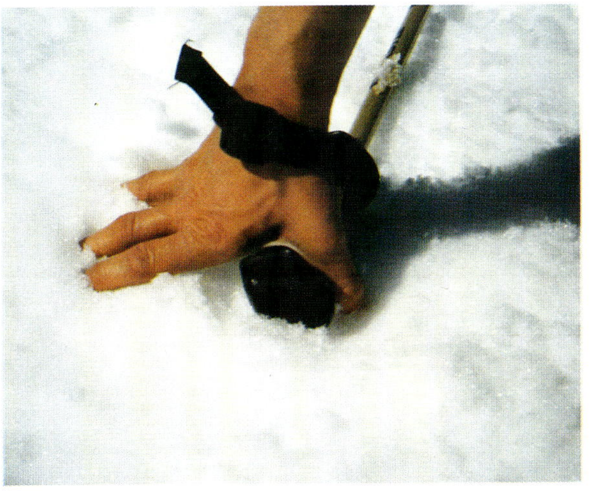

188 This injury usually results from a fall onto the hand when holding the ski pole, so that the ligament is stretched, producing metacarpophalangeal flexion.

Anatomical summary

The metacarpophalangeal joint of the thumb is a condyloid joint, which allows movement (flexion and extension) in one plane. In addition, it is capable of a certain degree of lateral movement and rotation, which assist the fine adaptation during the opposition of the thumb (Kapandji, 1966).

The head of the first metacarpal is spherical and fits into the glenoid cavity of the base of the proximal phalanx. This cavity is extended on the palmar side by the fibrocartilaginous palmar plate, in which are embedded the two sesamoid bones; into these are inserted the metacarpoglenoid ligaments (also called the accessory collateral ligaments). The insertion of the main collateral ligament is at the side of the head of the metacarpal, dorsal to the accessory ligaments. It extends towards the side of the base of the proximal phalanx, where it is attached near the insertion tubercle of the adductor of the thumb. Due to the shape of the head of the metacarpal and also to the point of proximal insertion of the main ligament, the latter is stretched when the metacarpophalangeal joint is flexed, and is relaxed when the joint is extended, while the accessory ligament is stretched during extension of the joint, and relaxed during flexion (**189**).

The stability of the metacarpophalangeal joint of the thumb is obtained firstly by the insertions of the adductor and the short flexor muscles into the sesamoid bone and its tubercle, at the base of the proximal phalanx, and, secondly, by the dorsal fibrous expansion, extending from the radial side in the form of the short abductor and from the cubital side in the form of the adductor. This fibrous expansion forms what is known as the dorsal capsule, which tightly envelopes the long extensor muscle of the thumb and slides forward with it when the joint is flexed, and backwards when it is extended.

The degree of possible flexion/extension varies enormously from one individual to another, and can range from 30 to 85°. As the mobility of the joint protects it from trauma, a joint which is not very mobile will be more liable to ligamentary injuries.

189

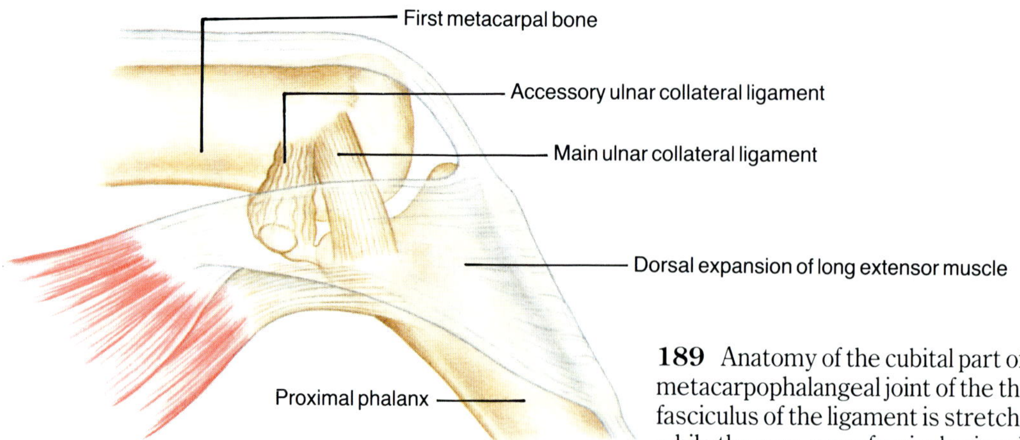

189 Anatomy of the cubital part of the metacarpophalangeal joint of the thumb: the main fasciculus of the ligament is stretched in flexion, while the accessory fasciculus is relaxed.

The mechanisms of injury

Ruptures of the cubital collateral ligament of the metacarpophalangeal joint of the thumb are generally considered to result from an abrupt and exaggerated abduction of the thumb. However, some authors claim that abduction with hyperextension is the mechanism because, when injuries of this type occur during skiing, hyperextension and abduction of the thumb against the ski pole occur, as the wrist is held back by the strap.

However, in reality the mechanism is quite different: whether considering a fall forward, landing with the thumb extended, or a fall when skiing, with ski pole in hand, the thumb, following the inertia of the movement, goes naturally into flexion at the metacarpophalangeal joint and into forced abduction. Stener (1962) demonstrates this in as anatomical study involving 42 fresh corpses and 39 patients who have undergone surgery; he stresses the frequent occurrence of a dislocation of the ligament, which is torn above the proximal edge of the dorsal capsule, thereby constituting a formal indication for surgery (**190**). This study shows that the dislocation of the ligament can be reproduced *post mortem* by a mechanism which combines the abduction, flexion and supination of the joint; this is the movement that occurs most frequently during a skier's fall with ski pole in hand, as is confirmed by Glas (1978) in a study concerning the morphological aspect of ligamentary injuries to this joint in skiing accidents.

Only when the main ligament is torn from its distal insertion, and the accessory ligament is intact, can it be said that the injury occurred in flexion. If both ligaments are injured (a rarer occurrence), one can conclude that the rupture occurred in strong abduction and probably in rectitude or light flexion, but never in hyperextension. In fact, although a dorsocubital capsular tear is very often found in association with a ligamentary injury, palmar injuries (tearing of the glenoid plate) are extremely rare. This observation confirms that the mechanism in question is definitely one of flexion–abduction.

190

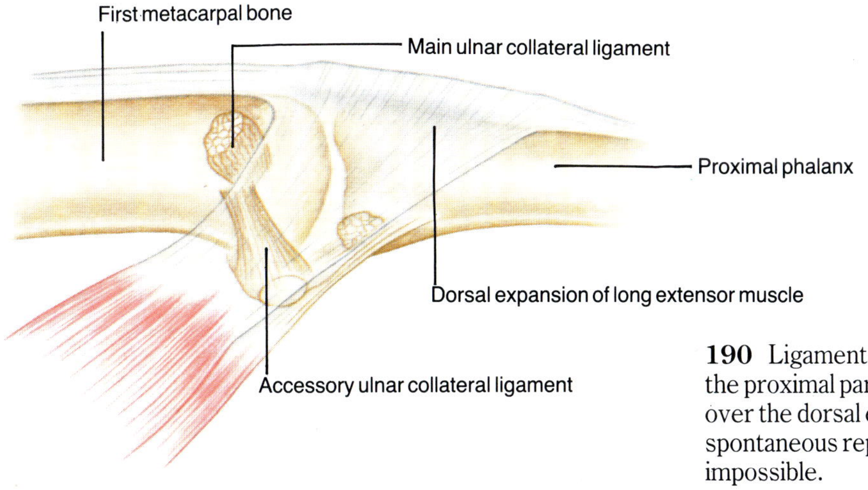

190 Ligamentary rupture occurs where the proximal part of the ligament is pushed over the dorsal extensor apparatus, so that spontaneous repair of the ligament is impossible.

Clinical and radiological examination

On examination there is usually pain on the cubital side of the metacarpophalangeal joint, with swelling and possibly an ecchymosis which may extend towards the thenar eminence, and a relative functional disability of the thumb, which has reduced strength.

When faced with this clinical picture, before any other type of examination is performed, standard X rays of the joint will determine whether the basal phalanx has been split (**191**). If the bone is split, but there is no displacement of the fragment, all that is required is immobilization in plaster for four weeks. However, if the detached fragment of bone is large, displaced or turned upside down, further examination is pointless, as this displaced fragment alone is sufficient indication that the main lateral ligament has been torn at its insertion, and that surgical treatment is mandatory. Widening of the joint space must be ruled out when no fracture is visible on the standard X rays (**192**). If this is the case, widening in valgus must be sought during both passive extension and flexion at 25°. Examination in passive extension is important as, in active extension, the dorsal capsule is tensed and provides a certain articular stability which can be misleading. Additional X ray examinations (passively held valgus or arthrography) do not provide sufficient diagnostic information for their use to be justified and, at times, they may even aggravate the injuries.

191

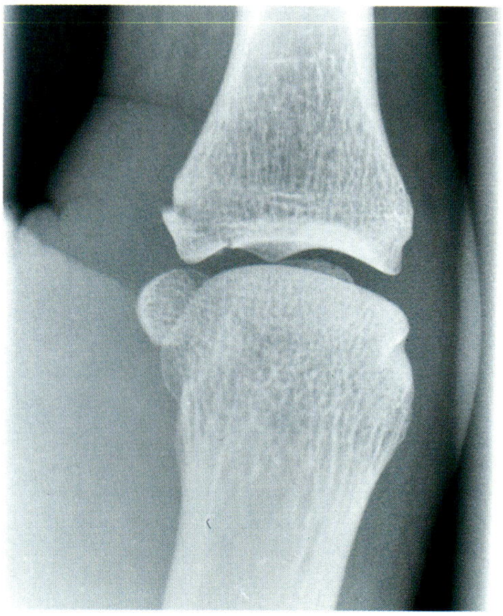

192

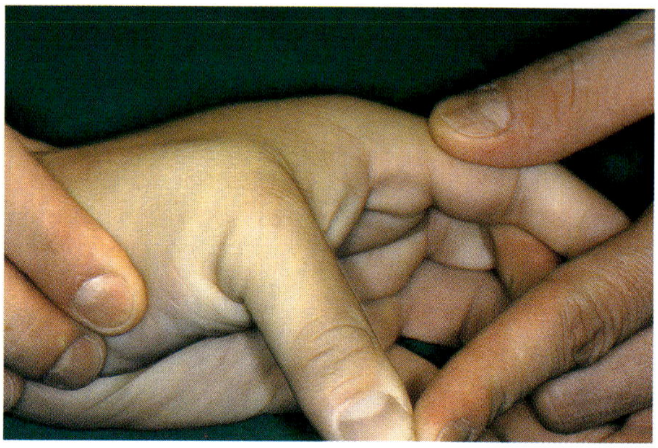

191 Undisplaced fracture of the distal attachment of the ligament; surgical treatment is not indicated.

192 Clinical instability of the metacarpophalangeal joint.

Indications for surgery

Surgery is definitely indicated where there is a bone chip that is displaced or turned upside down, and when (if no injury is visible radiologically) the widening of the joint is greater than the physiological widening of the healthy side. The physician must rely on this asymmetry, as it is impossible to give figures beyond which surgery is necessary; the normal abduction of the thumb varies between 15 and 40° in flexion, and between 5 and 30° in passive extension.

Many authors (Glas, 1978; Kapandji, 1966; Stener, 1962) have referred to the frequent occurrence of dislocation of the main ligament, torn above the proximal edge of the dorsal capsule – an injury which constitutes a definite indication for surgery.

Surgical technique

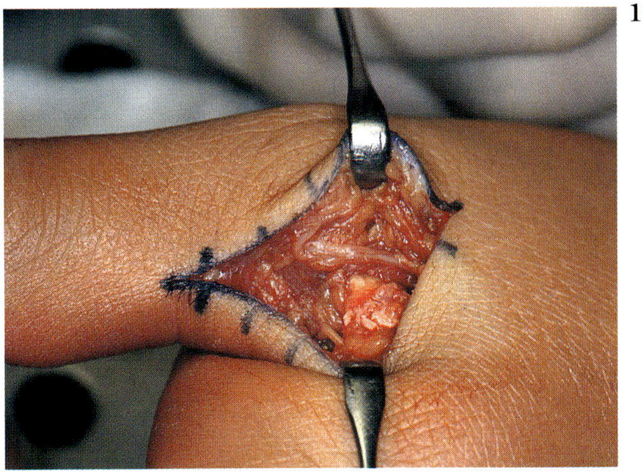

193

193 Operative view, showing the ruptured ligament and the displaced position of the proximal part of the ligament over the dorsal extensor apparatus.

An incision on the cubital side of the metacarpophalangeal joint reveals the sensory branch of the radial nerve, which lies dorsally. Above the edge of the dorsal capsule, one can sometimes observe a dislocation of the main ligament, torn distally (**193**). Once visible, the dorsal capsule is dissected longitudinally until the point of insertion of the main ligament is reached, at the base of the proximal phalanx. If the bone is split at the point of insertion of the main ligament, after reduction, the fragment of bone is fixed to the phalanx by means of a metal pin measuring 0.9mm; this pin passes obliquely through the whole base of the bone and emerges through the skin on the radial side. After a radiological check, the pin is cut near the fragment of bone on the cubital side and just under the skin on the radial side, where it is embedded to avoid any risk of infection. This technique enables the pin to be removed from the radial side, should rejection or infection occur, without tearing the fragment of bone, as may happen when it is removed from the cubital side.

One should always carefully rule out a capsular rupture of the dorsocubital part of the head of the metacarpal. This injury, which is common, must be sutured. When dealing with a torn ligament on a very small fragment of bone, the ligament must be sutured to its insertion, or to the neighbouring periosteum, if necessary. Fixing the ligament by means of a pull-out system through the proximal phalanx is not essential, as the thumb must, in any case, be immobilized. If the accessory ligament is also torn, generally at its proximal insertion, it must also be sutured. In order to do this, a non-absorbable suture of 4 or 5-0 should be used, as for the main ligament. Once the incision has been closed, layer by layer, the joint is immobilized for four weeks in a plaster cast, which holds the thumb away from the index finger. In the case of a split bone, the removal of the pin takes place immediately after the removal of the plaster, to allow active mobilization.

To conclude, it should be remembered that a thorough clinical examination generally makes it possible to estimate the seriousness of a metacarpophalangeal sprain of the thumb. The treatment must be adapted to the severity of the injury. Where surgery is indicated, the operation generally gives good or excellent results. This is not the case in secondary repairs, in which it is sometimes necessary to have recourse to a metacarpophalangeal arthrodesis, after attempted repair of the ligaments has failed.

Monoskiing and snowsurfing: new risks

Marc-Hervé Binet

The advent of new techniques of skiing is recent. It began with monoskiers, who started invading the powdery snow in the USA. In 1983, snowsurfers (or snowboarders) also began to appear in Europe. Today, surfing is a fully fledged sport (**194**). The number of surfers in France has risen exponentially: there were approximately 10,000 for the 87/88 season, and more than 30,000 in 88/89. There were probably over 100,000 surfers during the 89/90 winter season.

194

194 Snow surfing often involves spectacular acrobatics. *Courtesy of H. Tschan.*

Equipment

Monoskiing

The monoski (**195**) is a wide ski with two conventional bindings side by side. It is mainly used for skiing away from marked runs and in deep powdery snow. Monoskiing has been tried by about 5% of all downhill skiers, half of whom continue to practice it quite regularly.

195

195 The beginner often has difficulty controlling his speed on a monoski.

Snowsurfing

The equipment used in snowsurfing has evolved very quickly. The board, for example, which was originally shaped for use primarily in powdery snow, has become more sophisticated, with a similar shape to the monoski. Its hard edges make surfing possible on all types of snow and slopes. The main problems today concern the boots and the bindings.

Currently, there are two main options for the interface between foot and board: the first involves traditional ski boots, which are used by most beginners. In order to provide a low-cost product, retailers offer boards for rent which are compatible with ski boots. The binding consists of a system of plates, which cannot release in an accident. There are now boots designed especially for surfing. These are more flexible than ski boots and are used with the same type of binding. The second possibility is to use high boots, which are very flexible at the ankle. These are held in place by a system of bindings which, again, cannot release in an accident.

The position of the bindings on the board has undergone many changes since the birth of surfing. At first, the boots were placed almost perpendicular to the long axis of the surf board. Now the boots are placed at an angle of some 35–60° in relation to the long axis, with a greater angle of deviation for the rear foot (**196**). In surfing, a right-handed person is said to be in 'regular' position with the left foot forward, while a left-handed surfer is called a 'goofy-footer', as he surfs with the right foot forward.

Various disciplines have now developed within surfing, and it is practiced on either powder or hard snow, and both on and off marked ski runs. Following the example of downhill skiing, competition has developed on a national and an international level. Competitive surfing includes slalom (either standard or giant), moguls, and figures performed in a 'half pipe' (a huge semi-cylinder, cut into the ski run, which makes jumping easier). As well as having equipment especially designed for these disciplines, competitors often wear arm protection for slalom, and helmets when performing figures.

196

196 Today's bindings on surfing boards are not safe enough.

Accidents

Monoskiing

During the 85/86 season, the author investigated trauma due to monoskiing. The results showed several differences between monoskiing injuries and those occurring in downhill skiing, most notably a higher frequency of injuries to the upper rather than the lower limbs (the opposite is true in downhill skiing). However, monoskiers suffered relatively more ankle injuries than downhill skiers, and twice as many wounds.

Snowsurfing

A French study was conducted during the winter of 88/89 in an attempt to answer the following questions:

- Is there any type of equipment which is particularly dangerous?
- What is the actual relative risk compared with downhill skiing?
- What developments should be sought to improve the safety of surfing?

In order to answer these questions, three parallel studies were set up: first, a detailed survey of injuries compiled by doctors; second, questionnaires distributed by ski resort managers or sports shop owners to the surfing population; and third, a study of competitors in official competitions. From the results, the number of surfing injuries that occurred during the winter of 88/89 can be estimated at 423. Currently, this incidence is four times greater than that in downhill skiing. Not surprisingly, 50% of the accidents were found to have involved newcomers to snowsurfing, occurring within the first two days of their trying the sport.

These figures look more daunting than they really are. If they are compared with the downhill skiing accident rates from the time before the introduction of safety bindings, the risks involved in surfing may be viewed less pessimistically.

Mechanisms

Monoskiing

The following types of mechanisms are implicated in accidents:

Forward fall	32%
Forward fall with twisting	21%
Twisting	21%
Fall on the side	16%
Backward fall	10%

The release of the bindings occurs before the fall in 57% of accidents.

Snowsurfing

Accidents generally result from one of the following three mechanisms: (i) loss of control – it is difficult for a beginner to control his speed. Often he embarks upon a rapid descent, without being able to stop or turn at high speeds (**197**); (ii) collisions with obstacles – the winter of 1989 was particularly short of snow, and a significant number of accidents occurred through falls against tree stumps or stones, or against man-made obstacles, such as pylons and fences; (iii) leaving ski lifts – a significant safety risk in surfing is the position of the surfer on the board; the feet must be placed across its width. This position makes chair lifts particularly uncomfortable, and it is almost impossible to use drag lifts if both feet are attached to the board. Surfers using lifts with one boot detached are forced to leave the lift on one leg, making it very difficult to maintain balance.

Some significant differences were found in the localization of injuries in the disciplines of downhill skiing, monoskiing and surfing (**Table 4**). In surfing, the lower limb is generally more frequently injured, as it has no protection. Of the total number of leg injuries, 90% are to the lower leg. Some ankle injuries are specific to surfing; these include bone injuries (fractures of the malleolus and, in particular, of the talus) and ligamentary injuries (sprains of the tibiotarsal and the tibioperoneal ligaments). Talar fractures are also a specific surfing injury, and have been observed in more than 5% of lower limb injuries. Upper limb injuries are mostly fractures of the radius: surfers using their hands for balance may hit the ground with the heel of the hand.

If a comparison is made among the various types of injury in downhill skiing, monoskiing and surfing, there are no significant variations (**Table 5**).

197

197 Learners should always start surfing on a gentle slope. *Courtesy of R. Dahlquist.*

Table 4 Localization of injuries as a function of discipline.

	Lower limb	*Upper limb*	*Body*
Downhill skiing	36%	42%	22%
Monoskiing	34%	44%	22%
Surfing	65%	23%	12%

Prevention of injury

Monoskiing

A complete analysis of the results discussed above suggests particular safety measures which should be incorporated into the development of new equipment. These include the use of a 2–3m long, inelastic cord to stop the monoski if it comes off during a fall, and new bindings, in which the release of one automatically triggers the release of the other.

The author's study strongly suggests that manufacturers should try to develop a safer interface between foot and surfboard. A first vital step toward reducing injuries to the lower limbs will be the development of a way of releasing the boot – a safety measure which must be introduced as soon as possible, The criteria for well designed bindings include protection from twisting falls, resistance against axial pressure, and simultaneous release of both bindings. It is undoubtedly practically impossible to satisfy all these criteria at once. However, it is extremely important for beginners to have a system that allows the bindings to release whenever twisting forces are experienced.

Considering the significant number of accidents that occur during the first few hours of surfing, beginners, above all, should be aware of the following guidelines:

- Ski poles should be used on the first day; they avoid the risk of injury to the wrist during the first few descents.
- A short surfboard should be used; this makes surfing easier, and gives greater control. It also allows the surfer to roll the board in the event of a fall.
- The surfer should start on a very gentle slope; the technique of surfing is different from that of downhill skiing, and adaptation is more difficult that it is for monoskiing.
- The front boot must be adequately tightened; surfing necessitates a certain degree of flexibility at the level of the ankles and it is tempting, especially when using skiboots, not to tighten them completely. However, in order to control the board properly, avoiding inadvertent turns, it is necessary to fasten the boot properly.
- A safety leash should be used; this prevents both loss of the board and injury to a third party by a loose board.
- Both feet should be attached to the board when using chair lifts; it is still possible to keep both feet on the surfboard when taking a chair lift, thus making getting off easier. Ski lifts should be reversed for inexperienced surfers. Its is, of course, always possible to use cable cars.

Surfing is so appealing to skiers that, once they discover this new sport, many abandon traditional skiing. It is important to keep abreast of the developments in this discipline, if safety is to be improved. The author advises surfers and urges manufacturers to support research aimed at designing the safest equipment possible.

Table 5 Type of injury as a function of discipline.

	Sprain	*Fracture*	*Wound*	*Dislocation*	*Head injury*
Downhill skiing	37%	17%	10%	3.5%	3%
Monoskiing	32%	21%	20%	6%	4%
Surfing	45%	25%	8%	1%	4%

Mountaineering

Forces on the human body during falls on the rope and their consequences

Helmut Mägdefrau

As a result of the boom in rock climbing during the last ten years, there has also been an increase in the number of falls. Owing to the generally good belaying techniques, most rock climbing falls are harmless and do not result in acute injuries. However, to the author's knowledge, the last ten years have seen 13 accidents in the Alps, in which the impact force at the end of a fall has proved fatal; many of these were falls in which the climber was wearing just a seat harness, which only fits around the pelvis. Fractures of the spine were one of the causes of death, and may be simulated using a dummy (Schubert, 1984). An investigation in Spain (Martinez Villen, 1987) revealed two fractures of the spine (not fatal) as a result of falls while wearing only a seat harness. Whiplash and internal injuries, as well as headfirst falls, were further causes of death (Mägdefrau, 1987). In order to assess the risks of acute and chronic injuries, it was necessary to investigate the biomechanics of the human body during the impact at the end of a fall.

Forces on the body during a fall

Forces on the body resulting from falls onto the rope when belayed from above or on traverses are too weak to cause injury. Forces experienced in a fall when leading are much stronger, and have therefore been investigated in more detail (Mägdefrau, 1990). In this type of fall, the impact force is lowest when there is only one effective running belay, and is greatest when the rope friction on rock and running belays is extremely high. In both cases, the greater the ratio between length of fall and the dynamic rope length (i.e. roughly the greater the fall factor), the stronger the arresting force of the belay, and consequently the force exerted on the climber.

The forces were measured at the rope attachment point. A force of approximately 2.5 kN was registered at a fall factor of 0.5, using a single effective running belay with a rope turn-back of 165°, and single ropes and belaying with an automatic break (Antz patent). Most rock climbing falls lie within this range. At a fall factor of one, the force rises to about 3.5 kN (**198**). The other values on the curve were calculated using a theoretical force of approximately 4 kN at a fall factor of two. However, in the latter case, as no running belay is used (and therefore no rope friction is caused), the actual force on the body is only approximately 2.5 kN. Similar forces are found when using an Italian hitch (Munter hitch) or body plate. The use of a figure of eight descender for belaying is not recommended unless the device is clearly designed to use the smaller hole as a belay plate.

If there is significant rope friction over rock, the forces exerted on the body increase in direct proportion to the fall factor, as well as to the run out distance which is affected by the rope friction. Therefore, the fall factor and the length run out must be taken into consideration. With extreme rope friction, a fall factor of 0.4 and a length of fall of 4 m, the forces exerted on the body reach about 4 kN. With a fall factor of one and a length of free fall of 10 m, the force increases to 6.5 kN. However, such falls are rare, as few climbers would climb very far above the last running belay with such high friction on the rope.

The forces that arise are determined by the belaying methods and the natural friction; the weight of the falling climber is almost irrelevant. This means that light climbers are subjected to greater acceleration forces than heavy climbers. On the other hand, the impact pulse is accordingly shorter for lighter climbers than for heavier ones. By stating the impact force as a multiple of the force of gravity (g), the body weight can be virtually ignored for climbers with similar body proportions.

198

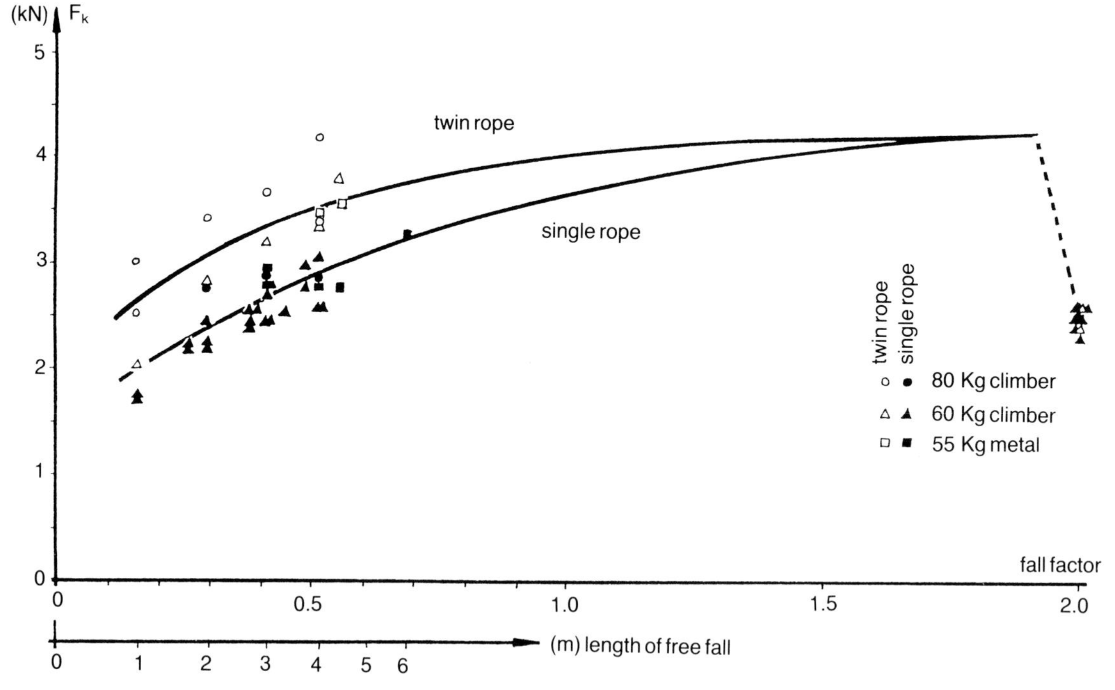

198 The force (F_K) exerted on the body at the rope attachment point related to the fall factor, with one effective running belay. The corresponding length of fall during the experiments is stated.

Consequences of the forces arising from a fall

In aviation medicine, the tolerable limits for upward acceleration are known for a person sitting tilted slightly backwards (Burton, 1985; Henzel, 1967; Webb, 1964). These investigations of the forces experienced during ejection from a fighter aircraft are very good for comparison with falls into a rope in an upright body position. It can follow that during the falls onto the rope, the limits for slight injuries are exceeded, if at all, only in extreme situations (**199**). No accident reports have been submitted up to the present time. If the body is kept almost upright, the impact produces mainly axial forces on the spine. The upright position is automatically achieved by wearing both chest and seat harnesses, i.e. with both chest and pelvis enclosed, as the body's center of mass is below the rope attachment point (**200**).

The situation is quite different if only a seat harness is used. In this case, the climber can achieve an upright sitting position during impact only if he manages to jump properly from the rock, or to catch the rope during the fall and thus right himself; in other words, he must control the fall (**201 & 202**).

199

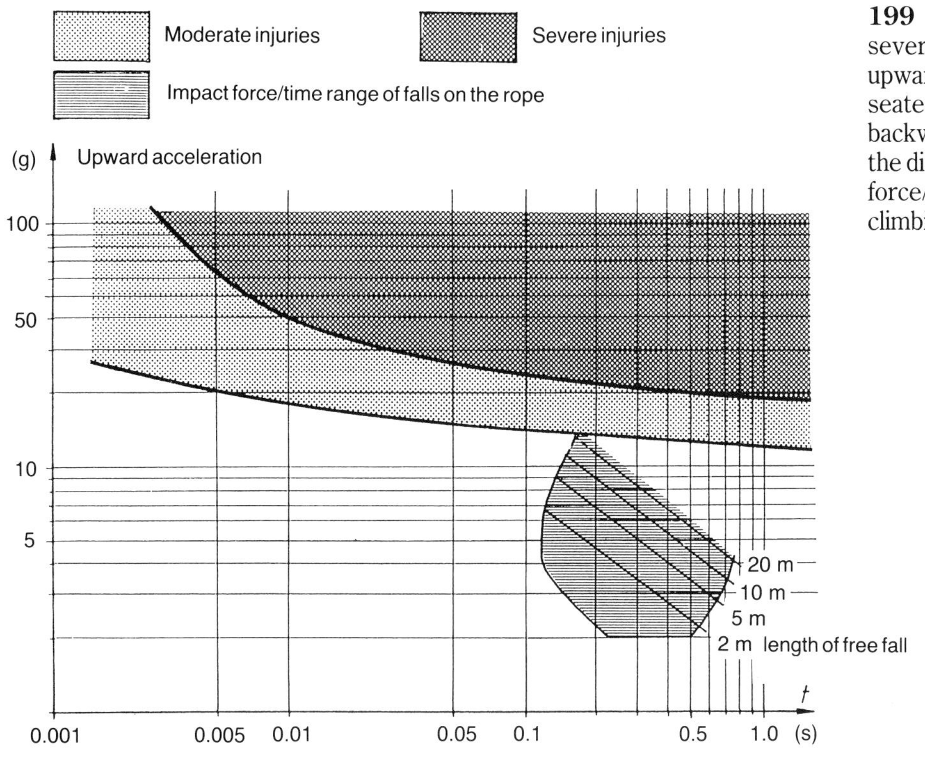

199 Regions of slight and severe injuries caused by upward acceleration when seated tilted slightly backwards. For comparison, the diagram shows the impact force/time region during climbing falls.

200

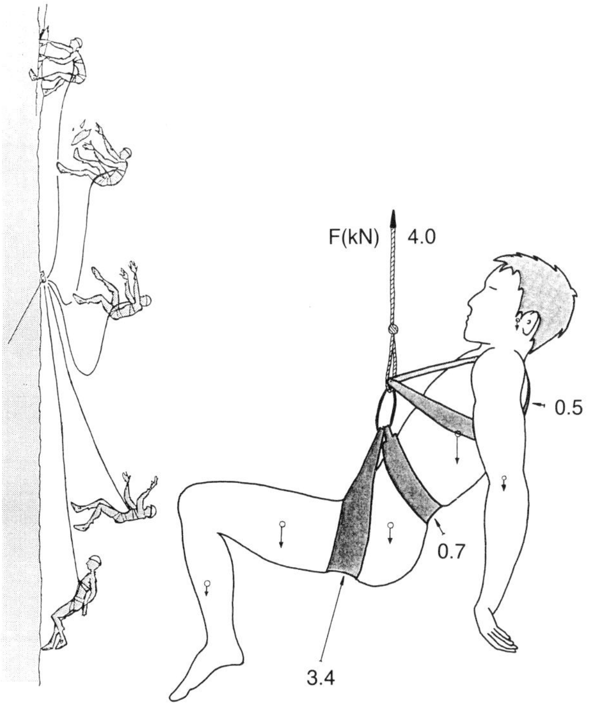

200 Uncontrolled fall with chest and seat harnesses. The body automatically assumes an upright, sitting position, because of the rope attachment point above the body's center of gravity. The forces produced and the centers of gravity of the limbs are shown in detail.

If the climber is not expecting the fall (for example, if a hold breaks loose), or if he is unconscious (perhaps because of rock fall), or if his body position is altered by hitting the rock, he cannot control the fall when wearing only a seat harness. Severe injuries may result, depending on the body position at the beginning of the impact: the spine is subjected not only to axial forces, but also to large bending forces; in addition, compression forces are exerted on the internal organs and whiplash injuries occur.

In the following cases, various body positions

of a climber wearing only a seat harness are examined during impact at the end of a fall, and the forces are determined for a well-trained climber.

In a horizontal position the abdominal muscles can withstand a dorsal (backward) bending moment of a maximum of 100 Nm. However, an impact force of 4 kN produces a bending moment of over 400 Nm on the waist belt of the seat harness (**203**). This explains the six known accidents with fatal spinal injuries.

201

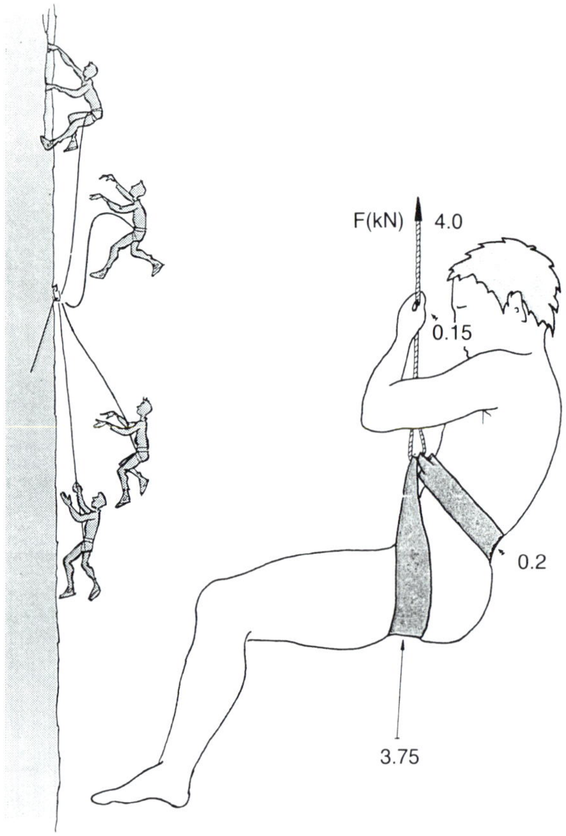

202

201 & 202 Controlled fall with a seat harness. By holding onto the rope, the body is prevented from tipping backwards. The forces produced are shown in detail. **202:** *Courtesy of D. Walser.*

In a horizontal, sideways position the internal organs, which are not protected by the pelvis or the spine, are subjected to extreme compression forces (**204**). This was the cause of two fatal accidents with rupture of the spleen.

If most of the centers of mass of the limbs lie on the same side of the rope during impact, only small bending forces occur, although large rotation moments can result, causing severe whiplash injuries (**205**). These injuries, which can be exacerbated by self-rotation of the body before impact, were the cause of two fatal accidents. The whiplash effect itself may cause the falling climber to hit the rock; in any case, he continues his fall in an upside-down, hanging position during the breaking distance, which is about a third to a half of the length of free fall. This position can result in head injuries before the fall has been completely stopped. In an upright position, only leg injuries are to be expected. The upside-down, hanging position can also aggravate head injuries before rescue, and was the cause of two fatal accidents and a number of severe injuries.

It is difficult to calculate how great the risk is of

chronic injury as a result of wearing only a seat harness. In the extreme cases described above, the muscles are stretched to their limits, even with an impact force of only 1 kN. Special training of the abdominal and vertebral muscles may raise the limits somewhat, but cannot prevent an excessive strain on the ligaments and even on the vertebral discs if the strain is unbalanced. Experiments to determine the strain limits were not possible because of the inherent risk of injury. In practice, several chronic injuries caused by repeated excess strain during impact are known. However, it does not necessarily follow that the impact force is the only possible cause, as previous injuries or damage for other reasons cannot be excluded.

203

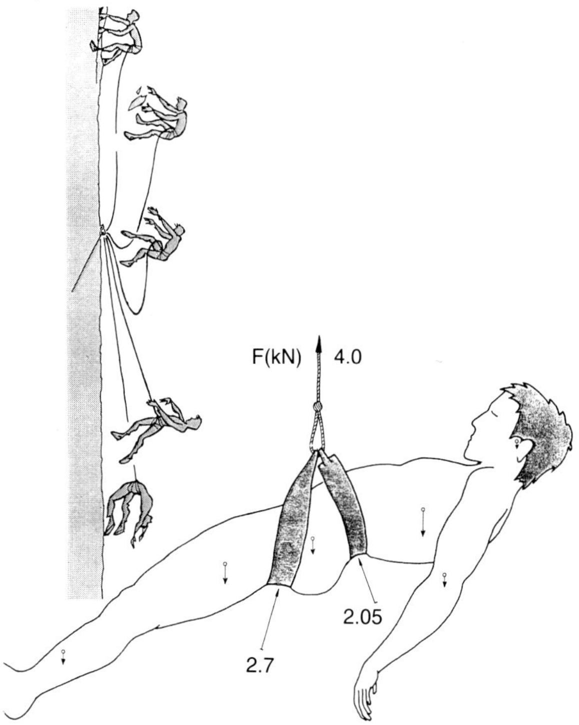

203 Uncontrolled fall resulting in a horizontal position at the first moment of impact. The forces produced and the centers of gravity of the limbs are shown in detail.

204

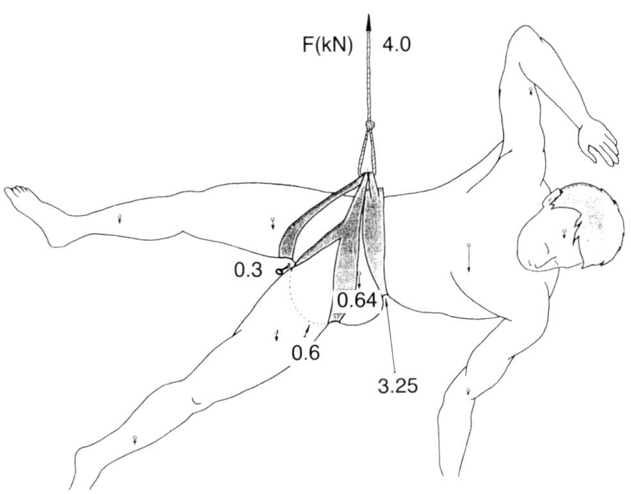

205

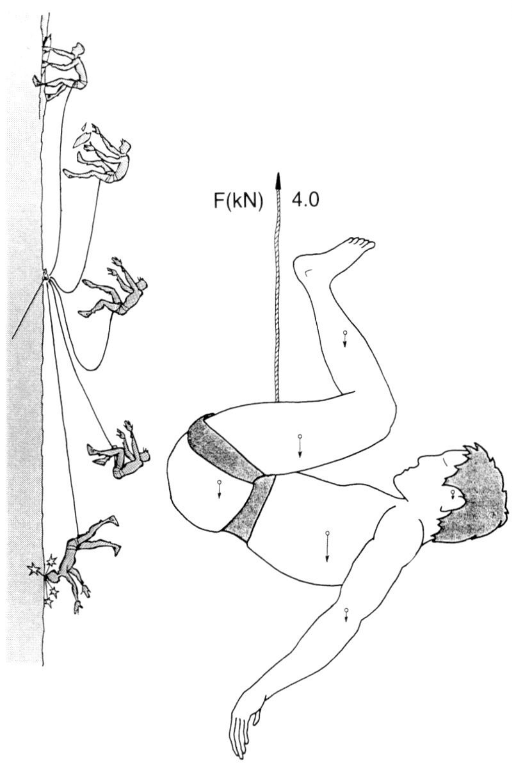

204 Impact forces in a horizontal, sideways position. The internal organs are subjected to a force of 3.25 kN at the waist belt of the seat harness.

205 Acute whiplash effect caused by an unbalanced distribution of the centers of gravity of the limbs, around the rope attachment point. Head injuries through hitting the rock are one possible consequence.

Conclusions

The only safe rope attachment methods that prevent both acute and chronic injuries are those using either a seat harness together with a chest harness, or a full body harness.

The frequent use by climbers of a seat harness on its own is due to the advantages it offers during climbing. When a climber wears only a seat harness, the rope drag (which can be as much as 0.25 kN) is taken up by the pelvis and is less strenuous for the shoulder and arm muscles. It is vital that climbers who prefer to have these advantages practice falling; a large number of falls can thus be controlled. The forces resulting from uncontrolled falls can be minimized by using very dynamic belaying methods (such as use of the body belay – it is vital that the belayer be securely anchored, preferably independently of the belay stance), twin ropes, and belays with low arresting forces. However, such methods have the disadvantage of a greater length of fall, and thus increase the danger of hitting either ledges or the ground; there still remains a small risk of acute and chronic injury from the impact force. Light climbers, in particular, should use a very dynamic belaying method, as they are subjected to greater forces than heavy climbers.

The only way out of this dilemma, between safety on the one hand and comfort on the other, is the use of the Erlacher patent (Mägdefrau, 1987); the rope fixed to the chest and seat harnesses is attached a second time to the seat harness, using the patented 'rope lifter'. The rope drag during climbing is thus transferred to the seat harness and greater comfort is achieved. During a fall, the rope is expelled from the 'rope lifter' and the falling climber, now properly attached with chest and seat harnesses, thus has ideal protection. Unfortunately, this method is only suitable for single, unforeseen falls. All fatal accidents caused by wearing only a seat harness occurred on routes where the 'rope lifter' could have been used. If several falls are to be expected, the use of the 'rope lifter' is too awkward.

Discussion

To date, there are no known cases outside alpine regions of fatal accidents caused by wearing only a sit harness. Only about 1% of all fatal climbing accidents occur in this way; the risk is therefore about as low as the risk of being struck by lightning when climbing. This is probably the reason why many rock climbers wear only a sit harness; they are prepared to take a small risk and, like most top athletes, ignore chronic injuries when trying to improve their performance.

In the author's view, it is necessary to differentiate between the various protection methods. If the different rope attachment and belaying methods are used appropriately on the various climbing routes, the risk when wearing only a sit harness can be reduced to an acceptable level. However, this assumes that the climber makes an individual decision before every route, which may be difficult for inexperienced climbers. For novices, distinctions between the various protection methods are useful and can be gleaned from instruction books. This is certainly better than receiving biased information from climbers who advocate wearing only a sit harness and who reject the current climbing theories as being incorrect and inapplicable on account of the low incidence of accidents. The large gap between theory and the practice shown in climbing magazines can only be overcome by proper distinction between the various rope attachment and belaying methods both in instruction books and in climbing magazines.

Head injuries

Jacques de Preux

Head injuries occur frequently during mountain climbing, and are also becoming an increasingly common skiing injury. The most important factor in head injuries is the early diagnosis of brain injury which requires neurosurgery.

A simple neurological examination to rule out mesencephalic compression (altered consciousness, pupils or hemiparesis) will lead to early referral to a neurosurgical unit for decompression treatment. The decision to transfer the patient must be taken as soon as possible, because there is an 80% mortality rate for intracranial hematomas that are not drained within four hours.

When examining a climber, it must be decided whether the patient has a lesion that requires immediate evacuation to a neurosurgical unit, such as a severe scalp laceration, an intracranial hematoma (extra- or subdural or intracerebral), a depressed skull fracture, or severely increased intracranial pressure due to cerebral edema (which requires cerebrospinal fluid (CSF) drainage, controlled hyperventilation or intracranial pressure monitoring).

The mesencephalon (**206**) is the most rostral part of the brain stem. In simplified terms, it could be said that it is the most sensitive part of the brain in neurotraumatology. Any expansive process (hematoma, edema) will displace the medial part of the temporal lobe, which will compress the mesencephalon at the tentorial incisure. The mesencephalon includes:

- The reticular formation, which is responsible for awakening. When injured, consciousness is altered.
- The nuclei of the third cranial nerves, which innervate the iris. When there is compression, an initial irritative stage (with pinpoint pupils) is observed, followed by a paralytic stage with mydriasis (homolateral to the compression), and finally, bilateral non-reactive mydriasis due to bilateral compression.
- The cerebral peduncles, through which the pyramidal pathways pass. Compression causes a contralateral motor deficit.

206

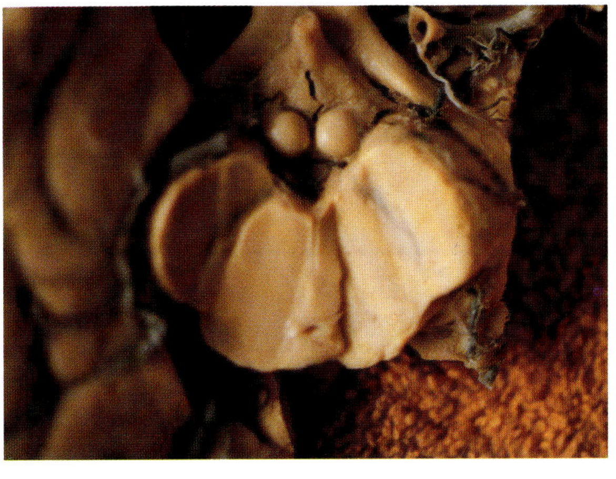

206 The mesencephalon, vital structure of the brain.

A lesion of any of these three structures must cause immediate alarm and be treated without delay, as irreversible damage or death may ensue.

At the site of the accident, a brief history of the circumstances of the accident should be taken. It should be determined, in particular, whether there was loss of consciousness (and, if so, its duration), amnesia (its duration before and after the head trauma must be specified) and any further loss of consciousness ('lucid interval'). It must be ensured that the patient has a patent airway and no respiratory distress (due to associated injuries, such as rib fractures and/or pneumothorax or aspiration), and is hemodynamically stable (shock is always due to associated injuries and not to head trauma). The head should then be examined for scalp lacerations or bruises, for depressed skull fractures (**207**) and for loss of blood and/or CSF through the ears, nose and mouth. The patient's state of consciousness, pupillary reactions and motor response to painful stimuli must be rapidly evaluated. The neurological evolution can be followed by determining the Glasgow score (which is the best ocular, verbal and motor response; *see* **Table 6**) every 15 minutes. The minimum score is 3, and 15 is a normal result. If the state of consciousness deteriorates or is abnormal for over one hour, or if one pupil starts to dilate or there are signs of hemiparesis, the patient must be transported to a neurosurgical unit as quickly as possible (**208** & **209**).

Emergency measures include controlling hemorrhage from scalp wounds (compression, tying off bleeding vessels), intravenous infusion for the correction of hypovolemia, endotracheal intubation and ventilation. Hyperventilation can relieve intracranial edema, as can the intravenous administration of a 12mg bolus of dexamethasone (followed by 3 × 4mg/24h), the infusion of 100ml of mannitol or the intravenous injection of 20mg of frusemide. Convulsions can be controlled by a 10mg intravenous bolus of diazepam or 1mg of clonazepam intravenously and 1mg intramuscularly, followed by 3 × 100mg/24h of phenobarbital. Thiamphenicol (2g/24h for 5–10 days) can be administered for antibiotic prophylaxis in open fractures or CSF leaks.

207

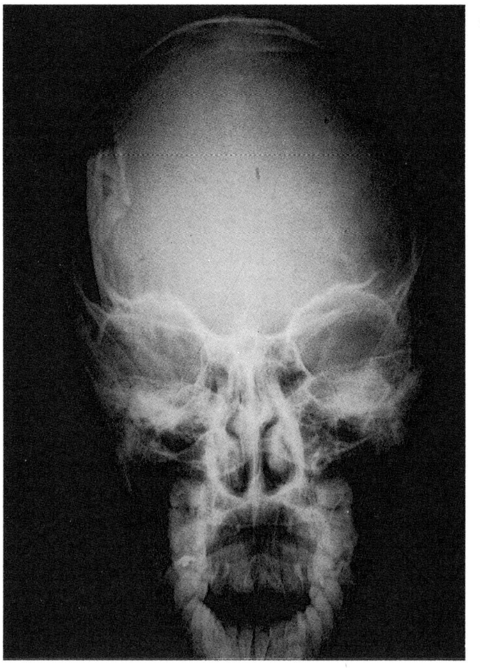

207 Compound depressed fracture of the skull.

Table 6 Glasgow coma scale.

Parameter	*Response*	*Score*
Eye opening	spontaneous	4
	on command	3
	to painful stimuli	2
	none	1
Best motor response	obeys commands	6
	localizes painful stimuli	5
	withdraws from painful stimuli	4
	abnormal flexion	3
	abnormal extension	2
	none	1
Best verbal response	answers appropriately	5
	confused answers	4
	inappropriate response	3
	unintelligible noises	2
	none	1

Patients should be transferred as soon as emergency airway and circulatory measures have been taken in the supine position, with the head raised to 30° above horizontal in order to lessen intracranial pressure.

208

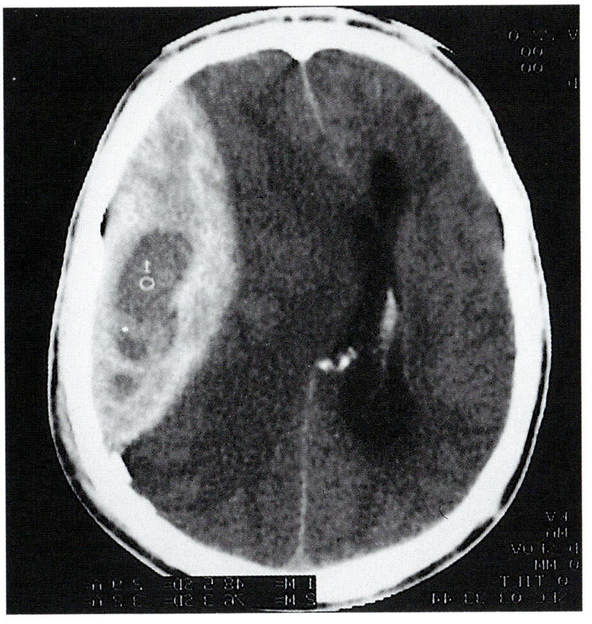

209

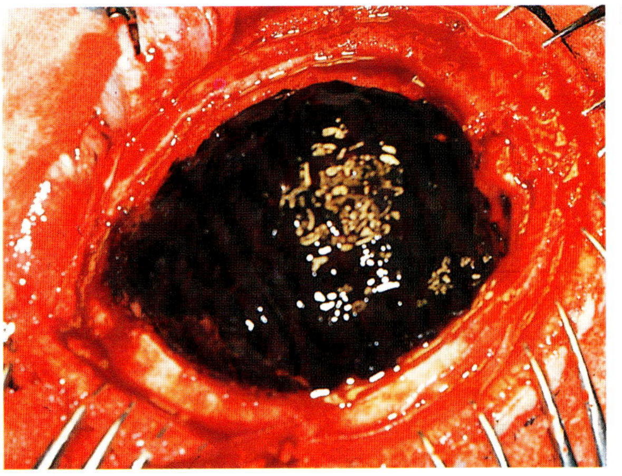

208 & 209 Epidural hematoma: **208** on the CT scan; **209** at operation.

Spinal injuries

Jacques de Preux

Spinal injuries requiring neurosurgery should always be suspected, so that the patient may be properly managed at the accident site, and transferred promptly to a hospital with a neurosurgical unit; in this way, further injury due to inappropriate manipulations of the patient may also be avoided. One must bear in mind that the neurosurgeon has, at most, 12 hours in which to relieve medullary compression.

Examination of the patient

As well as performing a neurological examination, the following injuries must be sought and ruled out:

- Unstable spinal fractures with a potential for further neurological injuries (**210** & **211**).
- Internal hemorrhage within the brain, the thorax or the abdomen (due to rupture of the liver, spleen or kidneys, for example).
- Hypoventilation due to fractured ribs, pneumothorax or aspiration.
- Hypovolemic shock.
- Limb fractures or injuries.

Furthermore, these patients often present with complications, which should be prevented from the start. These involve the skin (decubitus ulcers), the lungs (aspiration, pneumonia, poor clearing of secretions and sputum), the cardiovascular system (shock), the urinary tract (retention) and the digestive tract (ileus).

210

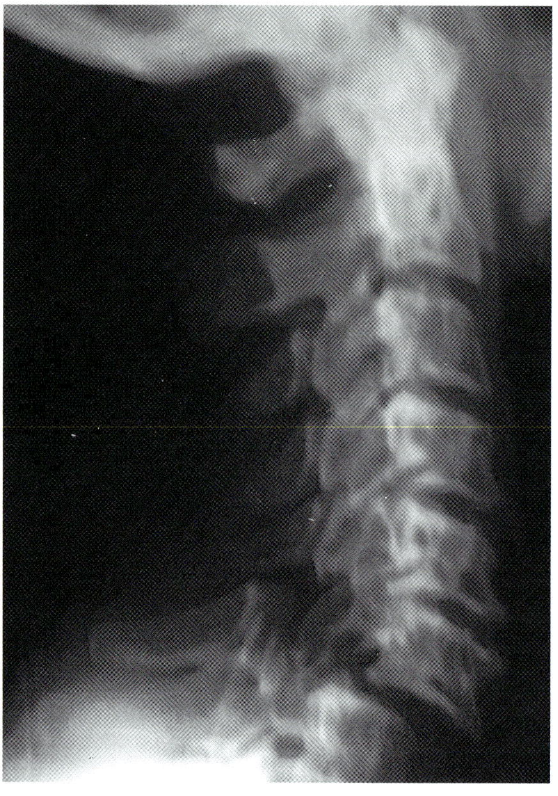

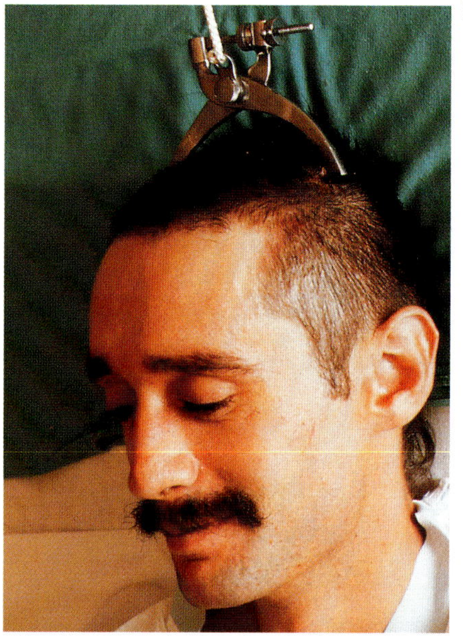

211

210 & 211 Fracture and dislocation of the cervical spine. Extension and reposition with Crutchfield tongs before surgical stabilization.

Fundamentals

The spinal canal is formed by the vertebral bodies, the lamina and the pedicles. These structures protect the spinal cord and nerve roots. The spinal cord extends from the first cervical vertebra (C1) to the second lumbar vertebra (L2), below which there are only spinal roots (cauda equina), and is essential in the following functions: movement, sensation, tendinous reflexes, breathing, urination, defecation, vascular tone, sexual function.

The cervical levels C1–C4 control respiration: innervation to the upper limbs extends from C5–T1, to the lower limbs from L1–S2 (**212 & 213**), and to the urogenital and anal areas from S3–S4.

When the spine is injured (fracture, dislocation, or both) the spinal canal is narrowed and can either compress or transect the spinal cord. Spinal cord or nerve root injury causes sensory and/or motor dysfunction of one or more limbs, with or without sphincter disturbances. When there are fractures of the neural arch or torn ligaments, the lesions are highly unstable and secondary displacement can cause further injury. When the patient presents with complete flaccid paralysis, the prognosis is often very poor. However, if some function is preserved or is recovered within 24 hours of injury, no matter how limited it may be, the potential for recovery is far better (**Table 7**). Nevertheless, the extent for recovery is highly variable and unpredictable.

212

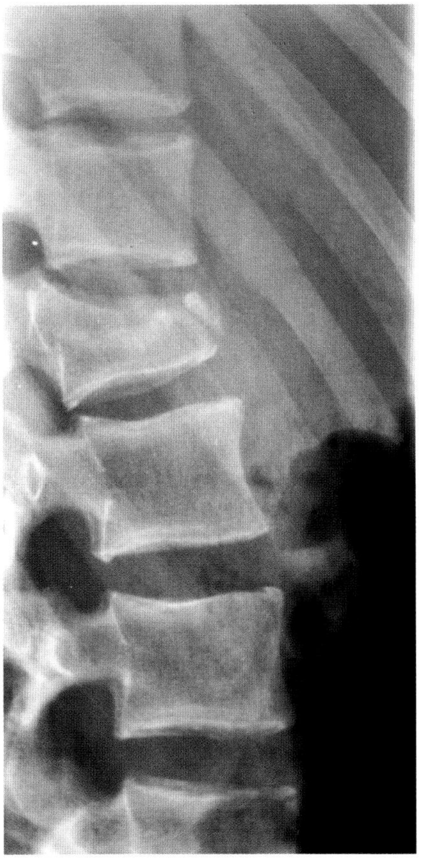

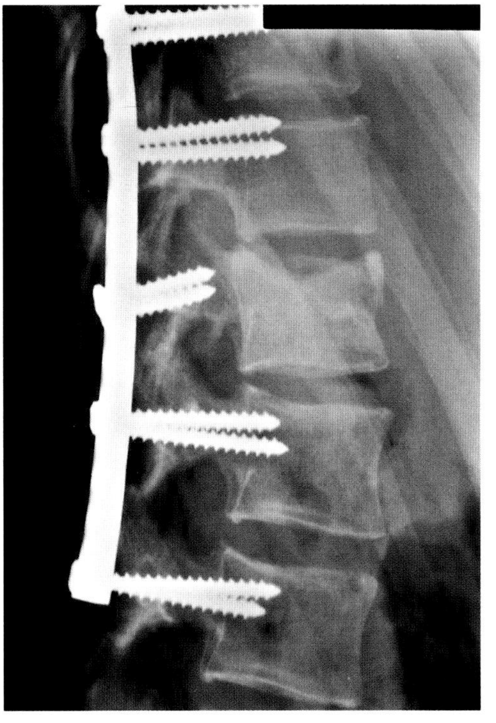

213

212 & 213 Burst fracture of the lumbar spine before and after surgical stabilization, with recovery of neurological function.

At the scene of the accident

It is important to find out from the patient (and from any witnesses) when the accident occurred, how it happened (how far did the patient fall, the type of ground fallen on, etc.), and whether there was an initial sensory or motor deficit, or if these developed subsequently.

Following this brief history, the patient must be examined **without the removal of any of his clothes**. The examiner must ensure that the patient's breathing is adequate (a high cervical lesion can cause paralysis of the diaphragm, and thoracic lesions can be associated with rib fractures and/or pneumothorax), that the airways are clear in an unconscious patient, and that the patient is hemodynamically stable. Thoracic or lumbar lesions can be associated with hemothorax or ruptured abdominal viscera. Autonomic dysfunction can cause vasoplegia and cardiovascular collapse.

Local examination (often difficult or impossible) seeks an electively tender spot or deformity, which often indicates the level of the injury.

The patient must then be examined for signs of serious spinal injury:

- Paralysis of one or more limbs or muscle groups (monoplegia, paraplegia, tetraplegia).
- Sensory deficits to either touch or painful stimuli, such as pinching the patient through the clothes; the patient should be given a brief, but thorough examination, from head to toe.
- Urogenital dysfunction (urinary retention, priapism).

First aid includes maintenance of a patent airway. If necessary, secretions or vomit should be suctioned. If a cervical lesion is suspected, intubation should only be performed by a competent person and the nasotracheal route preferred; the head should not be retroflexed. Circulation must also be maintained, with intravenous fluids if necessary. The patient should be installed on a firm, flat surface with a soft layer under the sacrum to avoid decubitus ulcers. If appropriate, immobilize the neck with sandbags or a stiff collar. If it is necessary to move the patient, this must be done by three or four people in unison, exerting axial traction on the head, and moving the patient as a rigid plank. No attempt should ever be made to reduce a spinal fracture at the scene of the accident.

Table 7 The most common lesions and their prognosis.

	Level	*Prognosis*
Cervical spine	odontoid	often good
	C5–7	recovery possible if early management
Thoracic spine	T4–5	very poor
	D12	recovery possible if early management
Lumbar spine	L1–2	recovery possible if early management

If available, these patients may benefit from a bolus intravenous injection of 12 mg of dexamethasone, from the infusion of 500 ml of dextran 70 (high molecular weight dextran) or from 20 mg of frusemide (intravenously).

Before transporting patients, make sure that they have patent airways and adequate circulation. Conscious patients without nausea are most easily transported on their back. Unconscious or nauseous patients should be transported on their side with a pillow under their head. Vacuum mattresses are the ideal solution.

Injuries in sport climbing

Anton Rañé and Conxita Leal

Sport climbing is a relatively young sport that has developed rapidly over the last 20 years. It began in the USA in the 70s and has since overtaken its mother sport, traditional climbing, in popularity and accomplishments. Much of sport climbing's appeal lies in the diversity of the challenges it offers, from cliffs to castles, and from the north face of a mountain to the facade of a bank. The sport has long exceeded any limits that were set for it (**214**). In addition, sport climbing has challenged the sporting norm and has opened up the world of competition; for the moment, it seems unstoppable and may even reach the Olympic Games.

Today's sport climber requires a high level of neuromuscular coordination, together with extreme sensitivity and strength in the fingers (**215**). The variety and intensity of demands made on joints, tendons and muscles in sport climbing have added a new area of pathology to those traditionally linked with climbing.

14

215

214 Muscular power is relatively easy to acquire with a few months of training, and the confidence that this sensation of 'traction power' gives the climber induces him or her to attempt increasingly difficult moves. The extreme strain placed on the osteotendinous structures in sport climbing inevitably increases their strength over a long period (three to five years), during which the capacity of the tendon to absorb great and repeated tractions progressively compensates for the lack of muscular capacity.

215 Maintenance of an ideal weight, and a long general warm-up (45 minutes to one hour), are two important factors in the prevention of overuse injuries. The stretches and mental relaxation should prepare the climber for the intense physical effort and the concentration necessary for the climb. A small amount of time should be allowed for recuperation after the warm-up, before beginning the climb. Climatic conditions should also be respected, and cold, in particular, must be avoided. Pictured here is Josep Battle, on the 'Ceguera temporal' Way, Montserrat, Spain. *Courtesy of M.A. Costa.*

Types of injuries

Injuries related to sport climbing (**Table 8**) can be divided into two main groups: those produced by acute trauma (i.e. a fall), which predominantly affect the bones and joints (fractures, dislocations, ruptures of ligaments) and which are classified as sporting accidents; and those that result from chronic overuse, due to the physical demands of climbing or training. Among the former, the most prominent are ankle and foot injuries, the fracture of the talus being one of the most frequent and the most feared because of its tendency towards aseptic necrosis. With respect to the injuries due to overuse, flexor digitorum tendinitis in the third and fourth fingers is the most recurrent and crippling injury.

Table 8 Most common climbing injuries.

Hand and forearm injuries	Flexor digitorum tendinitis of the third and fourth fingers Tenosynovitis: flexor carpi ulnaris and flexor digitorum Epitrochleitis, epicondylitis and radial styloiditis PIP joint capsulitis of the third and fourth fingers Spiroidal fractures of the proximal and middle phalanges of the third and fourth fingers Scaphoid and metacarpal fractures Avulsion fractures in the flexor digitorum attachment Tendon rupture of the flexor digitorum
Other injuries	<u>Bone fractures:</u> Talus Calcaneus Metatarsals Ankle Vertebral bodies, pelvis and skull <u>Tendon and ligament injuries:</u> Rupture of the Achilles tendon Flexor digitorum tendinitis of the feet Rupture of the tibial collateral ligament of the knee Patella attachment tendinitis Patella chondromalacia

Pathogenesis

The causes of the injuries are a function of the individual himself, as he goes beyond his own physical and mental limits. In the case of the overuse syndromes, there is a group of factors that may produce these injuries: great muscular development that has been acquired in a few months, as opposed to the very slow adaptation of the tendons to weights; irregular training or overtraining; absence of technical supervision; lack of or little warm-up; general diseases (respiratory diseases, abscesses); fatigue because of weak overall physical condition (maximal oxygen uptake of climbers with respect to other athletes), and also due to a lack of training and practice; insufficient fluid intake; unbalanced diet – in some, a hyper-protein animal diet is favored in an attempt to encourage development of the muscles, while in others a vegetarian diet is preferred, which, if the climber is not careful, may weaken him; and finally, faulty techniques – there are specific fingerhold positions that brutally strain the small structures of the interphalangeal joints and the flexor reflection pulleys.

The pathogenic hypothesis in these injuries comes from the pulley-tendon conflict at the proximal interphalangeal (PIP) level, including the repeated crushing of the tendon between bone and rock, added to the work due to the tension on this same tendon in the small holds (**216–218**).

216

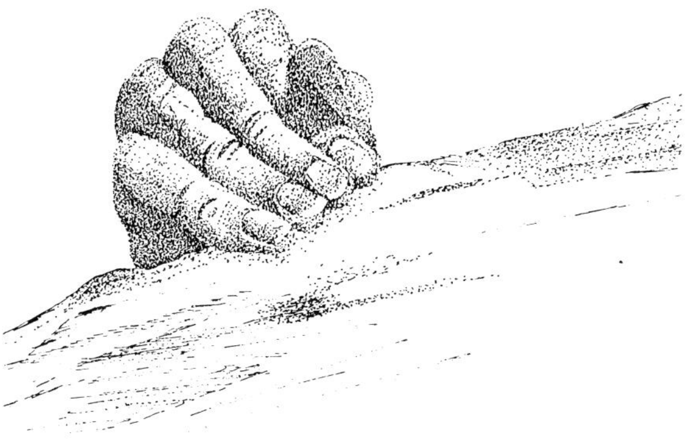

216 Shown here is a very common fingerhold position in small handgrips. This places considerable strain on the PIP joint and the flexor digitorum profundus and superficialis tendons, becoming more and more intensive with the increase in flexion. The physical demand in the DIP joint is smaller. This position should be avoided in training if possible.

217

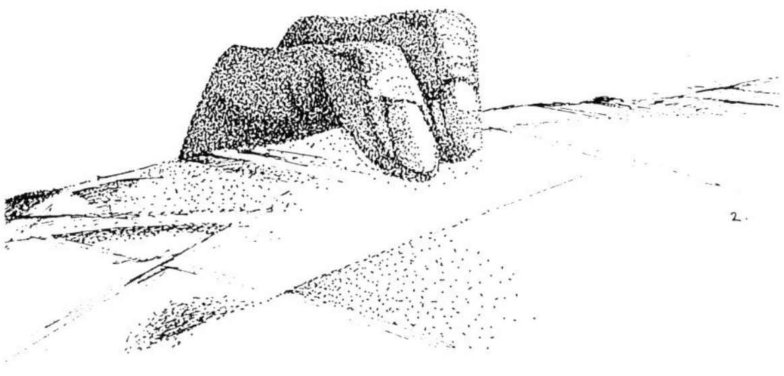

218

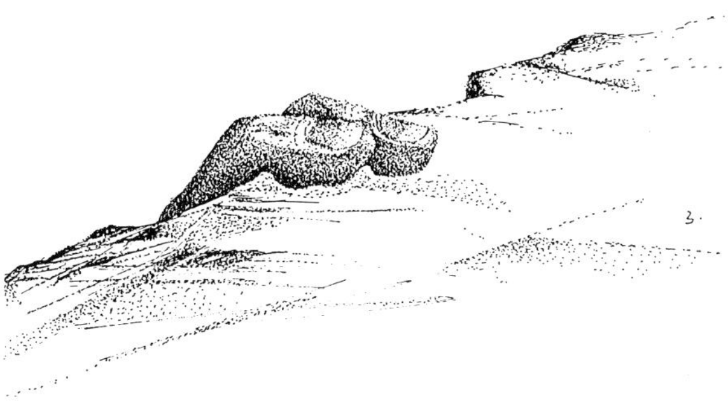

217 This position is less traumatic than the one shown in **216**. Strain on the DIP joint is minimal, but the joint and tendons of the PIP joint may be placed under the same amount of strain as in the previous position.

218 This is the most effective position for avoiding overuse injuries. Strain on the DIP joint is negligible, and the strain on the PIP joint and the flexor tendons' hold is about seven times less than in the previous positions.

Clinical features

Seventy per cent of climbing injuries affect the upper extremities and, in particular, the hand. The third and fourth fingers suffer the most injuries, especially the latter with 65–70% of the injuries. This predisposition towards the fourth finger may be explained by its frequent use as the sole supporting finger in very small holds. The distal phalanx has a smaller diameter than that of the second and third fingers. The symptoms of attachment tendinitis in the elbow and forearm are well known and similar to those produced by other sports.

Flexor digitorum tendinitis presents some peculiar characteristics: there may be an acute beginning (a crack) with pain and immediate loss of strength in the flexion of the affected finger; this pain may radiate out from the palm of the hand and from the inside of the forearm. There is generally an edema at the level of the proximal phalanx and the PIP. The resisted flexion of the distal phalanx over the middle one awakens a sharp pain. This injury may also appear more gradually, with an increase in tendon sensitivity during various training sessions, progressing to clear pain (**219–224**).

219

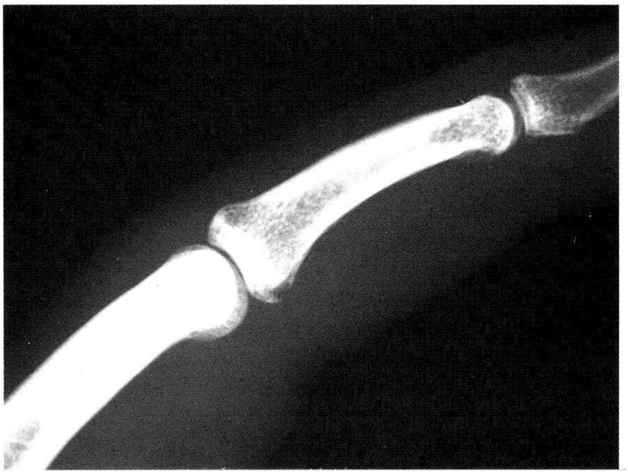

219 Bone avulsion fracture at the base of the middle phalanx, produced by a climber trying to contain a fall. This is not a common pathology, but it shows the great strain to which the small tendinous and ligamentary insertions are exposed.

220

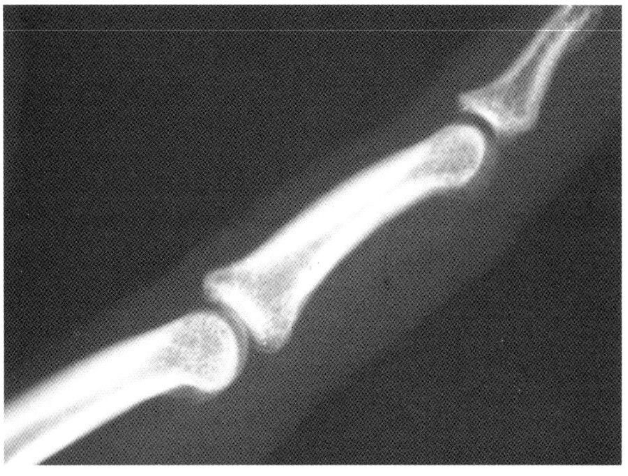

220 Great hypertrophy of the dorsal and palmar cortex of the phalanges, especially the middle one. This indicates the remodelling of these small bones, due to extreme biomechanical demands.

221 A normal subject for comparison.

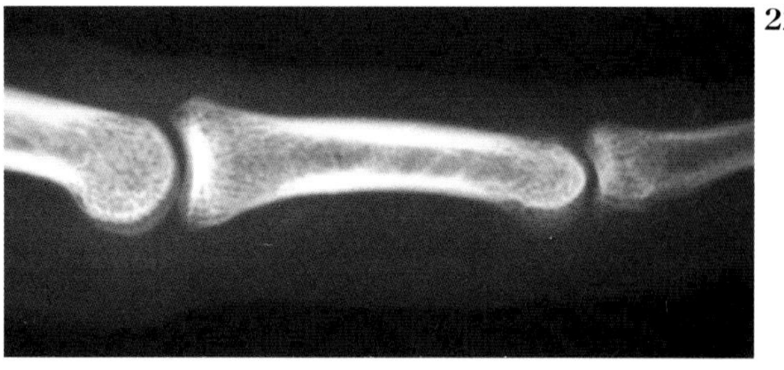

221

222

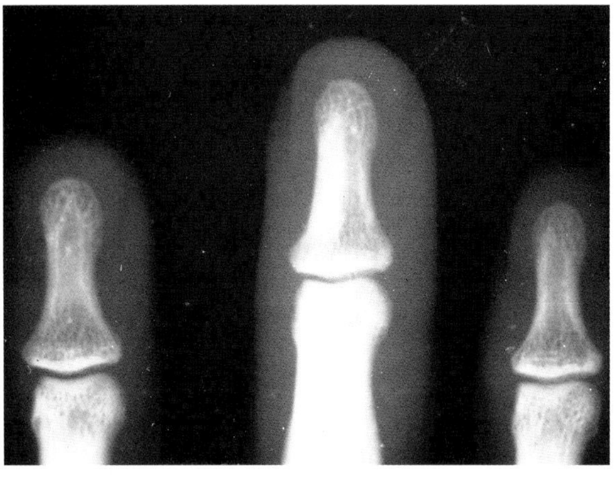

222 Degeneration of the distal phalanges with osteosclerosis, probably caused by repeated local ischemia, due to the compression of the fleshy part of the finger in the hold.

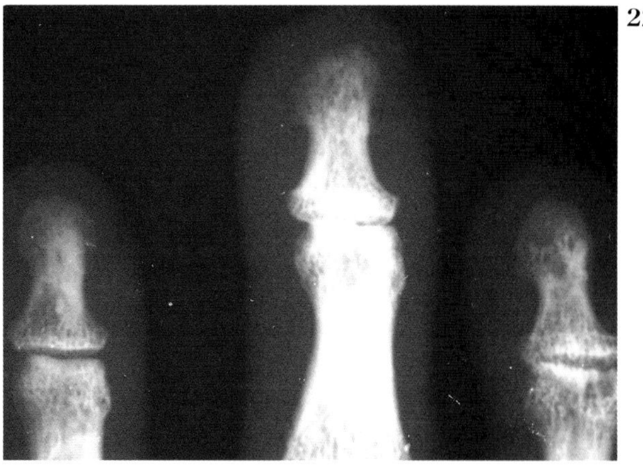

223

223 Sclerotic protuberances in the ungual apophysis.

224 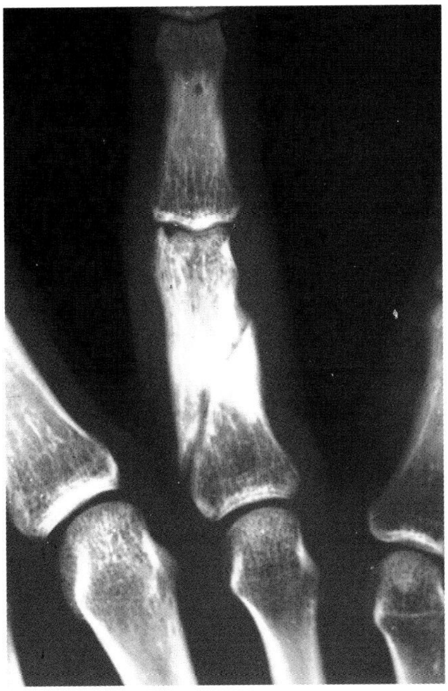

224 Oblique fracture in the proximal phalanx of the fourth finger, produced when the climber 'forced a path', hanging only from this finger in a hold.

Therapy

For injuries resulting from accidents, the usual treatment should be given. With respect to injuries caused by overuse, through the strain of training, one must contemplate multiple aspects, from the resolution of the local inflammatory symptoms, to the review and correction of the method of training itself, the equipment used, the general physical condition of the patient, diseases, and environmental factors in the development of the activity. Rotman, using the classification of strains that McKeag established, developed specific guidelines for treatment of this type of injury (**Table 9**).

All treatments should be adapted to the climber's personality. Generally, methods that are 'not too trying' and that do not remove him too far or too long from climbing are accepted. It is important to begin treatment immediately and with intensive medical–sporting supervision, until the symptoms disappear and the patient is completely cured.

Table 9 Clinical features and treatment of climbing related strains (according to Rotman, 1986).

Level	Clinical features and treatment
Level 1	Passing pain several hours after the activity (which may last one to two weeks). Increase in sensitivity. *Treatment:* Local cryotherapy Investigate training
Level 2	Long-term pain (two to three weeks) immediately after the exercise. *Treatment:* Local cryotherapy Reduce training by 10–25% Investigate technique, equipment and weather conditions
Level 3	Pain that lasts three to four weeks. It appears during climbing and gradually occurs progressively earlier in the activity. *Treatment:* Local cryotherapy Reduce training by 25–75%, do not train for a week Oral non-steroidal anti-inflammatory agents Specific physiotherapy
Level 4	Pain that lasts a month or longer. Before or at the beginning of the exercise. Greatly affects climbing. *Treatment:* Local cryotherapy Stop climbing Oral non-steroidal anti-inflammatory agents Specific physiotherapy

Prevention

As in other sports, there must be very careful general physical preparation to avoid strain injuries, but in climbing there are also several specific requirements that should be fulfilled. Before climbing, a complete warm-up should be performed, ending with finger stretches. Training by climbing seems to be less injurious than repeated traction sessions, which submit the tendon structures to weights up to the breaking limit. At-home training is more effective for the muscle system and less harmful to the tendons, tracting from wide holds (5cm or wider) and not from small 5 or 6mm grooves. As far as the muscle system is concerned, the repetition of unnecessary movement should be avoided, and a series of coherent exercises should be used instead, based on medical–sporting criteria that, until recently, have not been a part of the daily practice of most climbers (**225**).

225

225 In addition to a general physical activity, in which elasticity and neuromuscular equilibrium play a fundamental role, the sport climber must develop his or her own style of progression, in which his or her individual characteristics lead to a better development. Pictured here is Corinne Labrune, Troubat. *Courtesy of M.A. Costa.*

Training procedures and basic principles to avoid the risk of injury in modern free climbing

Lorenz Radlinger

Training producing a biological adaptation of tendons, ligaments and joints is basically feasible. The only problem is that it takes far longer than muscle development. Consequently, muscles can achieve what is required in climbing from a very early stage, but tendons cannot. The strength of the fingers must be built up through training in such a way that the risk of injury or wear and tear through constant repetition of a single finger exercise is minimal. A variation of exercises is necessary (**226**).

The exercises must be carried out progressively, and the critical exercises with straightened fingers in the general training program may be

226

227

226 Training at the board.

227 Strength training using the gripping machine.

228

229

230

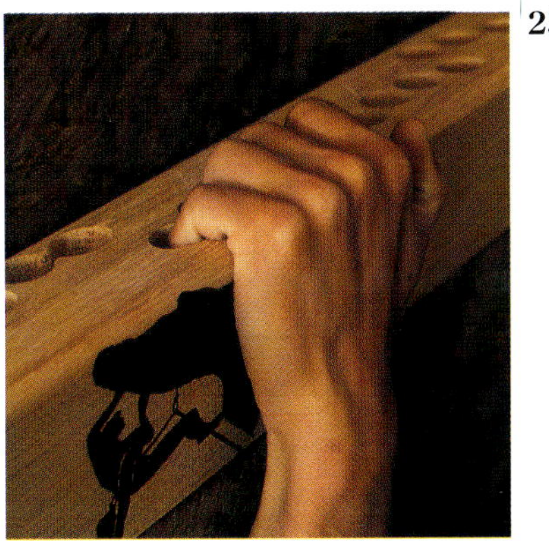

228–230 Different types of training may be used to attain maximal strength. Exercises must be progressive.

adapted to reduce the risk of injury. The exercises especially prepared to train for climbing have the same (or perhaps an even better) effect as constant training on increasingly smaller holds or grips. The advantage of using the gripping machine (**227**) in training lies in its ability to support the tendons along their sheaths. In this way, there is no jolting or bending of the joints. The training beams, in particular, are adapted to the anatomy of the individual and to the actual size of his hands and fingers. They can therefore replace training on small slopes and grips (**228–230**).

In spite of this, training to grip with the fingertips must be maintained using several variations because it is such an essential part of rock climbing. It is important for the biological adaptation of the joint (e.g. strengthening of the joint cartilages) that training also be carried out for the fingers. This is only possible with a sensible training load and program (**231**).

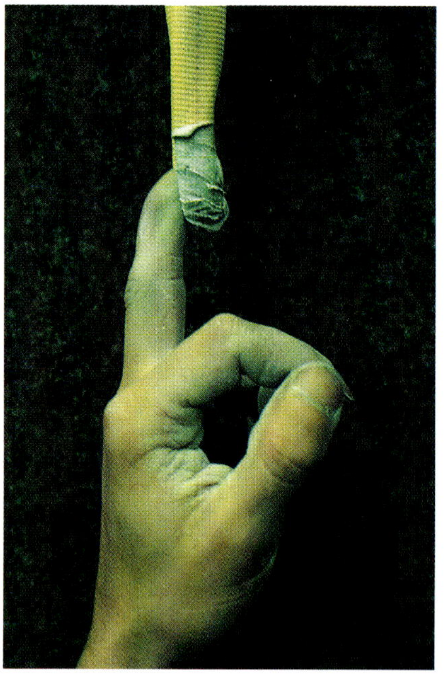

231

231 Single-finger exercise.

Basic guidelines for training and climbing

Warming up and stretching

Warming up and stretching should prepare the body for the exertions to come, and help it attain a higher level of performance. The warm-up should last about 10 minutes and include, for example, running, jumping, skipping, brisk walking, and climbing one or two easy routes. The intensity of the warm-up should be increased slowly during the 10 minutes. After warming up, the climber should stretch the muscles that are to be used in the climb. It is wise to carry out at least one or two exercises for the fingers and hands, shoulders, trunk and hips.

Strength training

Each new training exercise should be performed with care at the beginning, in order to learn the new techniques involved. The grips and exercises should be varied, so that the finger, hand, elbow and shoulder joints will not be put under constant strain by similar and unbalanced exercises. Between each series, the climber should actively loosen and relax the muscles that have been put under strain. He should not hold his breath, but should continue to breathe during periods of relaxation. Strength training should not be considered before the age of 14. It is begun with training to increase endurance, and continued, after some years of training, with combined and intramuscular training, and then with a 'periodization plan' (**232**).

Planning

The basis of any training is planning and organization, which must be systematic and aim at a particular goal, taking into consideration individual capabilities, and the ability to adapt to different conditions. The frequency and regularity of training, in particular, should be taken into account.

In both actual climbing and training, the frequency of training and, above all, the recovery periods allowed should be balanced (training units and pauses per week). Following a training or climbing session requiring intensive muscle endurance, 36–72 hours are needed for complete recovery. The length of the recovery periods will obviously depend on the amount of time spent in training, and on the level of training. Clearly, someone who is highly trained can recover far more quickly from intensive physical effort than an untrained person. Taking this into consideration, there should be at least two, and a maximum of five, training sessions per week.

During the course of a year, plans must be made for periods when no climbing, or less intensive climbing, is carried out (e.g. 4–6 weeks). In this way, a full recovery may be achieved after a long and intensive period of physical effort.

Like the warm-up, the 'cool-down' period plays an important role in all types of training, including training for climbing. One should run and climb sparingly, loosen up and stretch slowly and with ease. This brings the body back to its normal level of performance, speeding up and improving recovery.

Periodization plan: Stages in strength training in a program (20–22 weeks) designed to allow the attainment of individual top-level performance once in one year. **232**

<table>
<tr><td>Stage</td><td colspan="2">Preparatory period I</td><td colspan="2">Preparatory period II</td><td>Period of performance</td></tr>
<tr><td rowspan="2">Exercises</td><td colspan="2">Muscle development</td><td colspan="2">Coordination training</td><td rowspan="2">Endurance</td></tr>
<tr><td></td><td colspan="2">Combined training</td><td></td></tr>
<tr><td>Aims</td><td colspan="2">Muscular hypertrophy necessary for strength training</td><td colspan="2">Gradual transfer to maximal strength training</td><td>Adaptation to type of climbing</td></tr>
<tr><td>Time</td><td>8–10 weeks</td><td colspan="2">2–6 weeks</td><td>4 weeks</td><td>4 weeks</td></tr>
</table>

232 Strength (and the difficulty of the climbing) should be increased slowly and gradually. Top-level performance should be planned for on a long-term basis, so that it is reached in 6–10 years. Short-term (2–4 years) attempts are dangerous.

Muscular coordination training is suitable after a period of 3–5 years of build-up training (combined training after 2–4 years). It demands at least 3 training sessions per week with a training partner, under the supervision of a coach.

Further precautions

The climber should avoid getting cold; he should warm up well and wear gloves if necessary. During training for falls, the climber should get into the habit of not trying to grab onto something when falling (reflex grasping). He must quickly take his hands away from the rock face and fall in a controlled way (total security is, of course, absolutely necessary). Only then can he avoid the straining and tearing of muscle fibers and tendons, which often happens in these situations.

After training or climbing, the muscles should be massaged, and/or 'brushed' in alternate hot and cold baths. Care should be taken in taping, as symptoms of injury (e.g. tearing of tendon sheaths) will be masked, rather than remedies being found for the cause (e.g. incorrect movement, too much exertion). Apart from this, taping severely limits the levels of performance and freedom of movement. Training or climbing must, of course, be stopped when pain is felt. A new attempt may be made after an adequate period of rest. If the pain persists, it is advisable to consult a doctor or a physiotherapist.

Self-constructed climbing walls should not be equipped with grips that are too small. The grips should allow a beginner and an expert to climb. Their shape should be varied, to avoid monotonous types of gripping, and to make satisfactory endurance and technical training possible, rather than turning each climb on a climbing wall into strength training.

Passive pressure on the distal joints of the fingers should be avoided. The joints of the fingers are uni-axial hinge joints, i.e. they have a certain level of flexibility, and can only be moved in one direction. As a result, they react in a very sensitive fashion to sideways pressure, or turning and rotating movements. One should avoid sudden strains when snatching or jumping for a hold ('dynoing'), as well as high bursts of power during explosive climbing movements. During technical training, attempt large holds at first, followed by increasingly smaller holds, to learn precision of movement. Try, first of all, to make the hold at the highest possible point, or at the turning point of the movement, since the strength being activated at this moment is at its lowest level.

It must be kept in mind that there should be at least a 30–45 minute pause between the various routes to be climbed, if these routes are at the limit of the climber's capabilities. 'Top-rope' attempts should be avoided or drastically reduced. The climber should aim for regular 'on-sight' or 'red-point' ascents. The number of repetitions should be reduced. A healthy and gradual increase may be expected from the individual, physical, technical–tactical and mental capacities needed for climbing.

All the exercising procedures mentioned here have one aim: to avoid the risk of injury, especially among young climbers, who are at particular risk. If these guidelines are followed, the actual risk of injury is considerably reduced. It should also be remembered that all injuries heal either very slowly or, in some cases, not at all. In the latter case, there is an obvious decline in the level of performance. Therefore, if a sensible attitude towards climbing and training is shown, not only can the climber's level of performance be improved, but his health can be maintained at the same time.

Specific aspects of cave rescue

Olivier Moeschler

Unique aspects of cave exploration

Speleology presents specific and exceptional difficulties associated with the environment, as well as with modern methods of cave exploration.

Environment

Specific aspects of the underground environment are listed in **Table 10**.

Table 10 Specific aspects of the underground environment.

Total darkness
High humidity (95–100%)
Cold temperature (2–6°C in alpine caves)
Underground rivers, waterfalls and sumps
Confined passages (both horizontal and vertical)
Friable rock
Complex network caverns (e.g. Hölloch (CH, 1990): 133 km of galleries over only 7 km^2, with only two entrances)

Obstacles

- Vertical shafts have to be equipped with special static ropes, which are required to carry considerable quantities of equipment (**233**).
- Small vertical drops are free climbed, frequently over slippery and friable rock.
- Confined passages can be very complex (e.g. right-angled, S-shaped) and may have to be negotiated while the caver breathes out.
- Rivers require diving suits (**234**).
- Sumps necessitate special equipment adapted to cave diving.

233

233 The descent of a 40 m vertical pit on a fixed rope. *Courtesy of R. Wenger.*

234

234 Exploring an underground river. *Courtesy of R. Wenger.*

Characteristics of exploration

- Descents are usually performed by three to four cavers advancing individually and independently at intervals of several minutes, assuring their own safety.
- The return climb is far more difficult than the descent.
- The teams undergound are generally cut off completely from the outside world.

Accidents

Statistics

About two-thirds of the accidents in speleology are due to human error (Faust, 1983; Mallard, 1985) (**Table 11**). Accidents due to equipment failure are rare (8%).

Of interest are the facts that: two-thirds of the accidents occur in the entry zones (less than one hour from the cave entrance); the majority of the accidents occur during the return trip (Faust, 1983), and a considerable proportion of accidents occur during the physiological hours of sleep (Ritter, 1973).

Table 11 Mechanisms producing lesions in 143 cave accidents in France (1978–1981).

Falls	45%
Rockfalls	18%
Exhaustion	13%
Errors of judgement	8%
Drowning	5%
Miscellaneous	11%
	100%

Adapted and reproduced with permission from P. Faust: Accidents et incidents en spéléologie, Thèse de médecine, Nancy I, 1983

Types of accidents specific to speleology

Accidental suspension by rope

If loss of consciousness or extreme exhaustion occurs during a rope climb, the caver will remain suspended in the harness, tilted back to a variable degree. It has been shown that this type of suspension rapidly causes a state of collapse, the mechanism of which is complex and not yet fully understood (Amphoux *et al.*, 1984). Such a situation therefore constitutes an extreme emergency in which speedy intervention by fellow climbers can save lives.

Flash flooding

A sudden rise in the level of underground rivers can completely flood a gallery, or the water flow may be increased, rendering return impossible. The wait can be very long and uncomfortable (e.g. Diau, 1982: three cavers underground for seven days).

However, no deaths have been reported due either to hunger or to cold during prolonged waits (Marbach *et al.*, 1980). Fatal accidents occur when teams are surprised by the sudden swelling of a river. This may provoke frantic attempts to escape, with the cavers trying to dive while holding their breath (Vines, 1981).

Blockages in confined passages

In horizontal passages, problems often arise from badly executed maneuvers: panic and exhaustion may be induced without the victim really being trapped.

On the other hand, in accidents in vertical faults, cavers may find themselves wedged, particularly around the pelvis. Extraction is an especially delicate problem when the victim becomes stuck upside-down.

Problems during cave rescue

Duration of rescue operations

In contrast to mountain accidents, there are no short-cuts to reduce the length of underground operations: the rescue teams have to use the same route as that taken by the explorers.

Table 12 shows the duration of rescue operations up to hospitalization, in certain accidents that occurred in Switzerland between 1980 and 1987. In five out of 12 cases, the injured person went more than five hours without medical aid (maximum 20 hours). In five cases, admission to hospital was not possible until at least ten hours after the injury (maximum 68 hours).

Some rescue operations may take several days. It is therefore essential that fellow team members are trained to give first aid, and to provide protection from hypothermia by creating a 'favorable microclimate', while waiting for the rescue teams to arrive (Moeschler, 1982) (**235** & **236**).

235

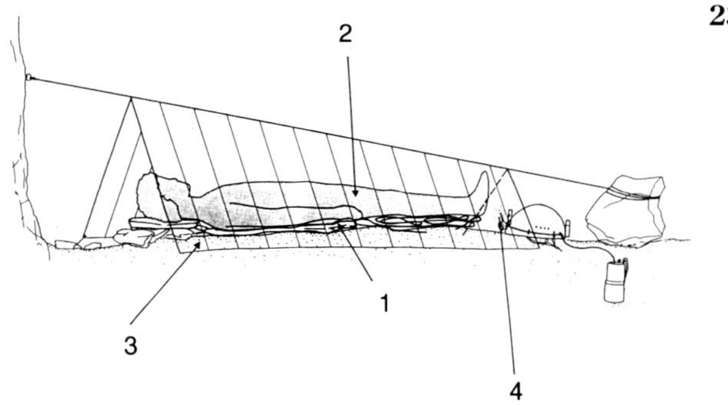

235 An improvised survival tent. Body insulated from the ground by a rope mattress (1), Patient wrapped in a rescue blanket (2), Tent made by a second rescue blanket (3), Tent warmed and dried out by carbide lamps (4).

Table 12 Time lapses (in hours) between accident and administration of first aid and evacuation, experienced by injured cavers in Switzerland (1980–1987).

Intervals between accident and arrival of help	*Intervals between accident and exit from the cave*	*Principal diagnosis*
3.5	6.5	severe head injury
5	12	facial injury
20	68	fracture of the femur
3.5	5.5	concussion
10	19	severe head injury*
2.5	5.5	fractured leg
2	3	exhaustion
5	11	bilateral fracture of the femur
2	4	mild hypothermia
3.5	8.5	cervical spine fracture
2.5	5	exhaustion
6.5	22	thoracic spine contusion

* died during rescue action

23

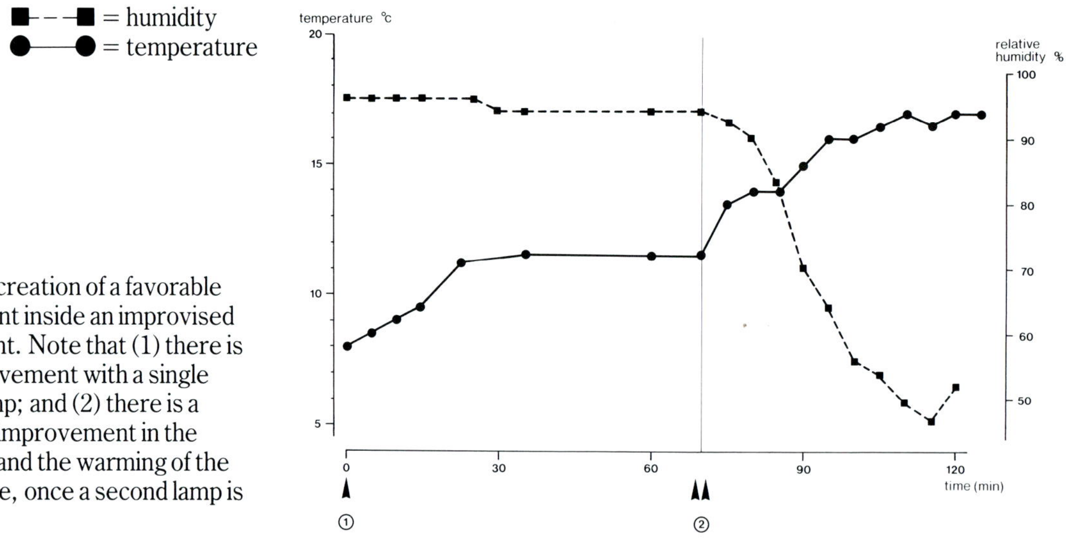

236 The creation of a favorable environment inside an improvised survival tent. Note that (1) there is little improvement with a single carbide lamp; and (2) there is a significant improvement in the drying out and the warming of the atmosphere, once a second lamp is added.

The doctor's role in cave rescue operations

Underground trauma can be defined as common injuries that are diagnosed and treated in the least desirable environment (Serra, 1984) (**237**). Working conditions for the doctor are made difficult because there is no possibility of immediate evacuation to hospital; vital functions, including core temperature, should be restored before underground transport; the injured person should be able to cope with change of position including vertical transport; the medical equipment available is very limited; and it is sometimes necessary to begin prophylactic measures normally reserved for hospitals.

237

237 The working conditions of a rescue doctor on the site of the accident, before the patient is transported elsewhere. *Courtesy of R. Wenger.*

Examples of medical problems encountered

- **Infection of open fractures/gas gangrene.** Open fractures and deep wounds must be treated on the spot.
  E.g. open fracture of the tibia, Cueto-Coventosa (Spain, 1985), depth 600 m.

delay before medical treatment	48 hours
duration of evacuation	72 hours
interval between accident and hospitalization	5 days

- **Fat embolism.** Fat embolisms are frequent in fractures of long bones in young adults.
  E.g. compound fracture of the femur, Sieben-Hengste (Switzerland, 1981), depth 400 m.

delay before medical treatment	20 hours
duration of evacuation	48 hours
interval between accident and hospitalization	3 days

- **Acute renal failure.**
  Acute renal failure is due to prolonged states of shock, acute urinary retention, and crush syndrome.
  E.g. fracture of the lumbar spine, paraplegia, Sotano de San Augustin (Mexico, 1980), depth 570 m.

delay before medical treatment	60 hours
duration of evacuation	81 hours
interval between accident and hospitalization	6 days

Problems of underground transport

Protection of the injured person

Underground evacuation can be a very long, difficult and extremely laborious process (**238**) and, in spite of precautions, the injured person is often exposed to: numerous shocks from the rock walls and the ground; transport in a vertical position; spray and waterfalls, and the cold. Thus it is essential that the victim is well prepared for the evacuation.

- **The stretcher:** several kinds of stretcher exist that are specially designed for underground conditions. The 'Spéléo-Secours Suisse' has developed one that protects and immobilizes the patient, even during transport in the vertical position (Probst, 1982) (**239**).

238

238 Transport of the stretcher into a gallery.
Courtesy of Y. Grossenbacher.

239

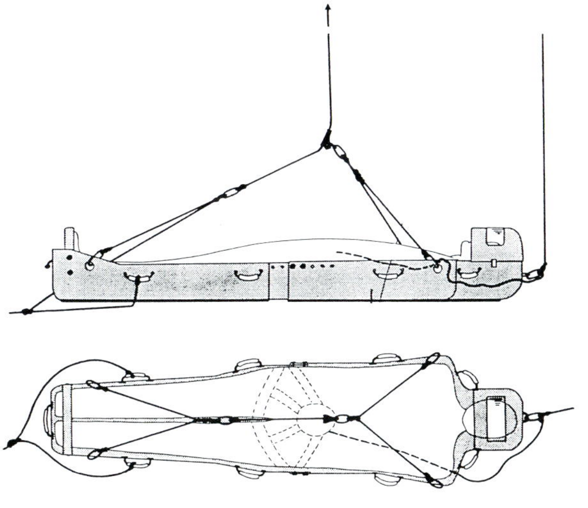

239 Cave rescue stretcher of the Schwarzer type (Spéléo-Secours Suisse), which provides the injured person with the best possible protection.

- **Means of immobilization:** these include strapping the victim in a parachute-type harness which allows vertical transport without sagging of the body (**240**); immobilizing the whole body in a split cast of expanded polyurethane foam; skin traction kits for injuries of the lower limbs, and cervical collars and inflatable splints when indicated.
- **Protection against the cold:** protection is provided by the use of special 'Holofill suits' (**241**), offering easy access to any part of the body, and developed by the 'Spéléo-Secours Francais'; waterproof bags in damp caves, and warming bags on the thorax.

240

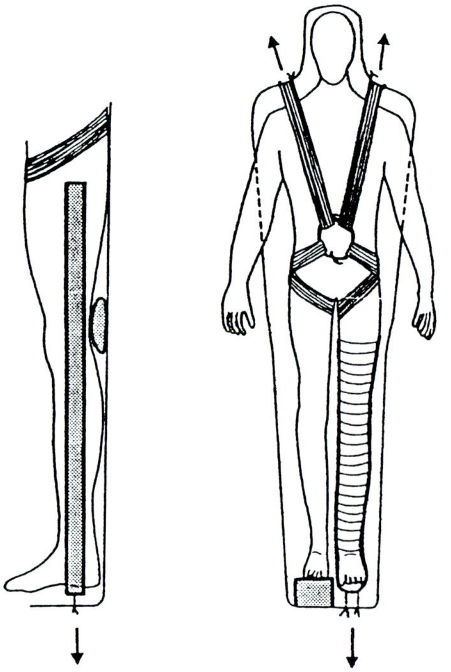

240 The patient is attached to the inside of the stretcher, allowing a fractured femur to be placed in extension.

241

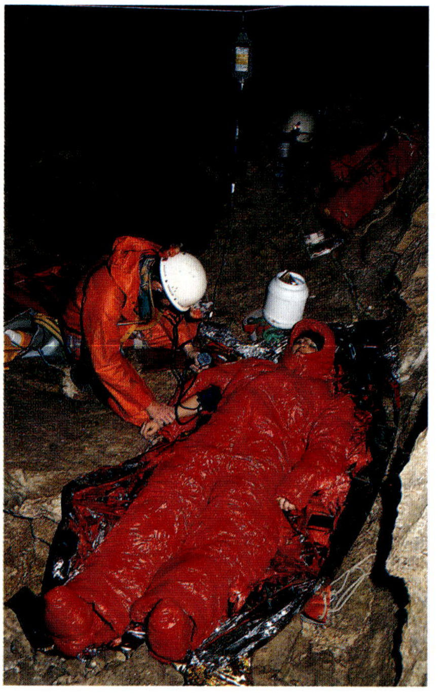

241 The injured may be protected from the cold by being placed in a special 'Holofill suit' (designed by the Spéléo-Secours Français). *Courtesy of R. Wenger.*

Problems of vertical transport

In certain shafts, with the careful installation of equipment, the stretcher can be raised in the horizontal position (**242**). However, in most situations, the stretcher has to be hoisted in a vertical position for long periods (e.g. rescue in the Z 49, Switzerland, 1981; shaft 150m deep.)

The vertical position can create several problems:

- **Hypotension** may be caused by several factors (particularly in patients with unstable vital signs), including venous stasis in the legs, aggravated by straps hindering venous return; reduction of the thoracic pump effect due to suspension in a harness; and muscular inactivity. These phenomena can be significantly reduced by restoring an adequate circulating volume, bandaging the legs for venous support, putting the stretcher into the vertical position very slowly, and transporting the patient as much as possible in a semi-inclined position adjusted by a traction rope (**243**). Transport in a truly vertical position is rarely necessary for a prolonged period.

- **Rotation of the stretcher and collision with the walls** may occur in wide bell-shaped shafts. Rotation, which may cause very serious motion sickness and the danger of bronchoaspiration in the totally immobilized victim, can be prevented by hoisting the stretcher into the semi-inclined position, thus allowing better mechanical coupling, and by someone accompanying the stretcher (**244**).
- **Venous stasis and edema of the lower limbs.** To prevent ischemic complications due to edema during vertical transport, circular splinting systems should not be used in cave rescue operations.
- **Sagging of the victim** must be avoided, particularly in cases of leg or pelvis fracture. Good fixation in a parachute-type harness is essential.

242

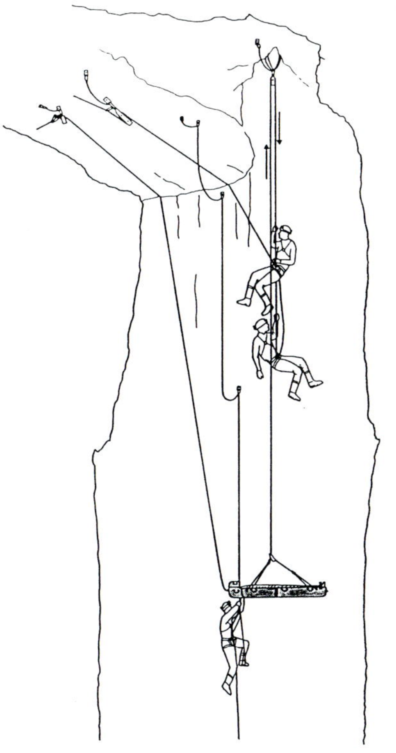

242 The patient is hoisted using a system based on counterweights: two rescuers act as weights and regulate the speed, while a third accompanies the stretcher using an independent, fixed rope.

243

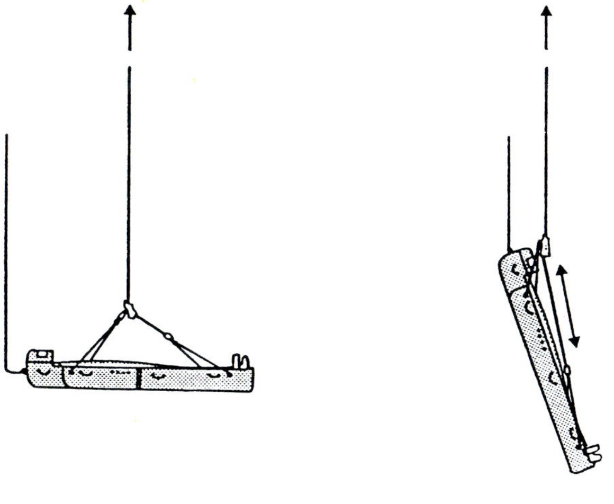

244

243 Where feasible, i.e. if the pit is broad enough, the stretcher is hoisted up in a semi-inclined or even a horizontal position.

244 Difficult maneuver at the top of a pit. *Courtesy of R. Wenger.*

Narrow passages

Confined passages, through which an ordinary stretcher is too large to pass, may be very difficult to negotiate. In some cases, the victim may have to be taken out of the litter to 'fit through the squeeze' (**245**).

245

245 Narrow passages may be very difficult to overcome. *Courtesy of R. Wenger.*

Conclusion

Cave rescue presents particular and unusual difficulties that require the use of specific equipment and techniques, and the selection of experienced and well-trained rescue teams.

Because of the usually lengthy and very difficult nature of cave rescues, it is necessary for a physician to stay with the patient throughout the operation. It is of some consolation that the victims are most often understanding, and accept responsibility for their overall care.

Proposal for a modular first aid kit for climbers, mountain guides and climbing doctors

Urs Wiget

The proposed modular system contains a **basic first aid kit** for alpinists without specific medical knowledge, and **extension modules** for mountain guides and climbing doctors (**246–249**).

The content of a first aid kit will largely depend on the quality of instruction that the user has

Module 1 Basic kit to be used by non-guide and non-physician climbers		**Quantity**
Pain	Paracetamol – ASA – codeine tablets	(10)
Cramps	Hyoscine–N–butyl bromide tablets	(5)
Fever	Paracetamol – ASA – codeine tablets	
Cold	Xylometazoline nose drops	(1)
Cough	Dihydrocodeine 25 mg capsules	(5)
Diarrhoea	Loperamide capsules	(5)
Vomiting, motion sickness	Metoclopramide 10 mg tablets	(5)
Throat	Small tablets to suck	(10)
Eyes	Oxybuprocaine **0.2%** salve (tubes)	(1)
Lips	Phenol – zinc oxide salve (tubes)	(1)
Disinfectant	Povidone-iodine 10 ml	(1)
Dressings	2 compressed gauze bandages 5 cm × 10 m, adhesive plaster, some disposable adhesive dressings, skin closure strips	
Miscellaneous	1 small pair of tweezers, 3 scalpel blades, alcohol preparation pads	
	Instructions for use (includes prescribed dosages)	
		Module 1 = 300 g (9 × 17.5 × 3 cm aluminum box)
Module 2 Extension module for well-instructed mountain guides		
Antacid	Aluminum and magnesium carbonate tablets	(10)
Angina pectoris	Nifedipine 10 mg capsules	(4)
Altitude exhaustion	Acetazolamide 500 mg capsules	(5)
	Prolintane 10 mg and multivitamin tablets	(10)
For sleep	Midazolam 15 mg tablets	(5)
Ampoules	Tramadol 100 mg	(2)
	Water-soluble prednisolone 250 mg	(1)
Miscellaneous	1 disposable syringe 2 ml, 3 injection needles, alcohol preparation pads	
	Instructions for use	
		Modules 1 + 2 = 370 g (10 × 22 × 3 cm aluminum box)
Module 3 Extension module for climbing doctors		
Altitude	Dexamethasone 4 mg tablets	(20)
Antibiotics	For example: co-trimoxazole 160/800 mg and/or cefadroxil 1000 mg tablets	(5)
Ampoules	Adrenaline 1 mg	(1)
Wounds	Atraumatic thread and straight needle for sutures without needle holder	
		Modules 2 + 3 = 380 g

246 Modular extensible first aid kit: (1) the basic kit to be used by non-guide and non-physician climbers; (2) this extension module is added to (1) for well instructed mountain guides; (3) extension to be added for climbing doctors.

received. For example, mountain guides equipped with extension modules take compulsory medical instruction courses at least once a year, and the basic kit and modules are regulated and changed if necessary. Once the guides have satisfied certain conditions, they are allowed to include injections in their kits.

Proposals for first aid kits are obviously subjective, but, based on the author's 13 years' experience working with over 300 active mountain guides, the following criteria are suggested for consideration:

- **Usefulness.** When an accident occurs in the Alps, mountain guides sometimes find that climbing doctors in the area have no useful medical equipment, a point which must be taken into consideration when putting together a first aid kit for guides.
- **Weight.** The container in which the kit is carried must be very light, yet sufficiently strong to protect the contents, even if a climber sits on it. The author has found sheet aluminum of 1 mm thickness to be the best material.
- **Size.** Dressings can present a problem because of the space they take up in a kit. However, by teaching guides to improvise dressings and bandages, the problem can be avoided. Equipment such as scissors can easily be replaced with scalpel blades.
- **Type of medication.** The list should include stable galenic preparations with maximum therapeutic security, and should not contain narcotics.

247 Basic first aid kit for climbers without specific medical knowledge.

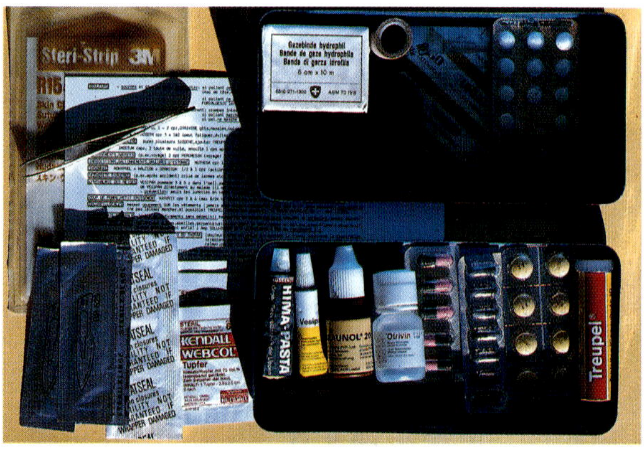

247

248
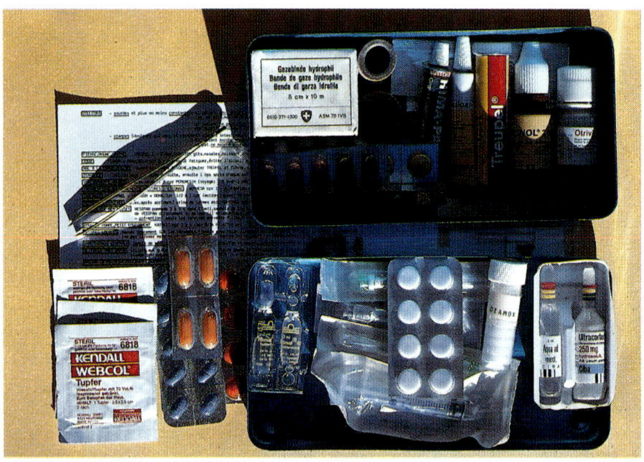

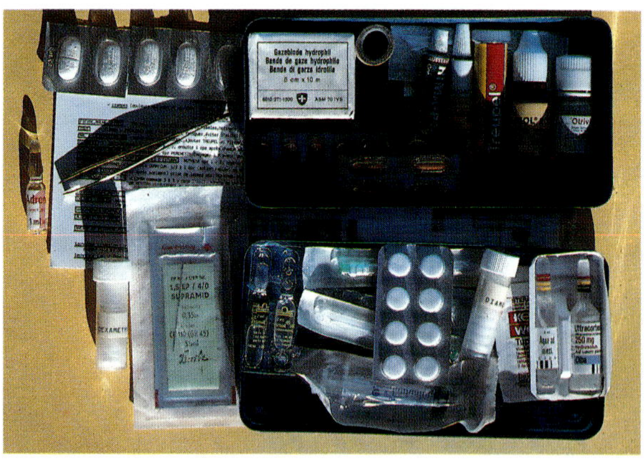

249

248 Basic first aid kit with extension module for mountain guides who follow medical instruction courses at least once a year.

249 Basic first aid kit with extension module for mountain guides, including extension module for climbing doctors.

Proposal for a first aid rucksack for mountain rescue doctors

Mountain rescue doctors who operate in the mountains on an independent basis cannot always count on particularly sophisticated equipment being available on board rescue helicopters or in ambulances, and they may have to rely on a first aid rucksack (**250** & **251**).

Depending on the amount of liquid in the kit, the rucksack may weigh 12–17 kg, and the doctor may be obliged to carry the rucksack on his back, even skiing with it. The following is a suggested contents list:

- **For reanimation:** intubation kit with tubes and spare batteries, ventilation unit with masks, oxygen bottle and Mayo tubes, aspiration unit for adults and children, 102 flask (1 L/200 bars, fixed flow 4 Lmin), thoracic drainage kit, Heimlich valve, gastric catheters.
- **For monitoring:** portable cardioscope with electrodes and spare batteries, head lamp with spare batteries, stethoscope and sphygmomanometer, urinary catheters, lubricating and anaesthetic gel.
- **Infusion:** various venflons and butterfly valves, adhesive silk, 2–3 × 500 ml sodium chloride, 1–3 × 500 ml dextran.
- **Drugs:** nifedipine 10 mg, nose drops, eye salve, aerosol spray of salbutamol, dexamethasone 4 mg.
- **Ampoules:** etomidate, midazolam, suxamethonium, pancuronium, ketamine, atropine 0.5 mg, adrenaline 1 mg, morphine 10 mg, naloxone 0.4 mg/ml, droperidol 2.5 mg, flumazenil, pentazocine, verapamil, glucose 40%, lignocaine 1%, frusemide, sodium bicarbonate 8.4%, dexamethasone 40 mg, phytomenadione, methylergometrine, hyoscine–metamizole, deslanoside, propanolol, haloperidol, fentanyl.
- **Miscellaneous:** dressings, fixation devices (leg, cervical spinal column), disinfectant, syringes and needles, tourniquet, tweezers, scissors, needle holder, atraumatic thread, surgical gloves and sterile sheets, scalpel blades, mylar survival blankets ('space' blankets), thermal bags, knife, pencil, note book, matches, thermos flask, case history sheets, protective clothing (e.g. gloves, balaclava, 'Goretex' suit), climbing gear (e.g. harness, carabiners).

250 

250 First aid rucksack for rescue physicians.

 251

251 An open first aid rucksack for physicians.

Proposal for a first aid kit for those trekking without a doctor

(based on four persons trekking for four weeks)

For those trekking without a doctor, the content of a first aid kit will depend upon the medical knowledge of the group. During treks beyond the easy reach of professional rescue outfits, a simple broken arm or dislocated shoulder can either lead to a tragedy or be managed easily by the layman if he has been previously trained. Training in medical procedures before the trip is as essential as the choice of equipment. It is also recommended that all trekking groups should have an advising doctor at home, who is aware of the different problems that may occur at high altitude.

Training before a trip should cover: the use of drugs (including antibiotics); intravenous and intramuscular injections; splints with casting material; management of dislocations (e.g. shoulder, finger); simple suturing with disposable materials; sterilization of instruments in the field, and how to improvise in the field.

The basic kit should include the following:

Pain	Tramadol 50 mg capsules	60
Fever/pain	Paracetamol – ASA – codeine tablets	200
Cramps	Hyoscine–N–butyl bromide tablets	40
Inflammation	Ibuprofen 600 mg tablets	40
Cough	Dihydrocodeine 25 mg capsules	50
Antacid	Aluminium and magnesium carbonate tablets	100
Vomiting, motion sickness	Metoclopramide 10 mg tablets	40
Diarrhoea	Loperamide 2 mg capsules	60
Laxative	Bisacodyl 5 mg tablets	20
Throat	Tablets to suck	100
Nose	Decongestant nose salve (tubes)	2
	Tablets for steam inhalation	50
Antihistamine	Astemizole 10 mg tablets	30
For sleep	Midazolam 15 mg tablets	30
Eyes	Oxybuprocaine 0.2% salve (tubes)	2
	Atropine – antibiotic drops	2
	Neomycin + bacitracin salve (native population)	10
Dermatology	Simple corticoid salve (tubes)	2
	Corticoid + antibiotic + fungicidal salve (tubes)	2
Teeth	Temporary filling cement	2
Antibiotics	Co-trimoxazole 160/800 mg tablets	50
	Cefadroxil 1000 mg tablets	20
Altitude	Acetazolamide 250 mg tablets	40
	Dexamethasone 4 mg tablets	20
Ampoules	Tramadol 100 mg	10
	Adrenaline 1 mg	2
	Water-soluble prednisolone 250 mg + solvent	2
	Disposable syringes 5 ml	15
	Injection needles	30
	Alcohol preparation pads for disinfection	100
Surgery	Atraumatic needle + thread	4
	Sterile tweezers, forceps, scissors, needle holder	1
	Sterile scalpel blades	20
	Sterile skin-closure strips	20
	Sterile latex glove (pairs)	8
	Fiberglass casting material 7.5 cm × 10 cm	3
	Material to pad under cast 8 × 10 cm	4
	Disinfectant in crystal form to dissolve in water	

Dressings	Elastic bandages (large)	3
	Elastic adhesive bandages 6–8–10 cm	2
	Sterile gauze 8 × 12 cm (boxes)	4
	Antibiotic gauze, 10 pieces (boxes)	3
	Selection of disposable adhesive dressings	
Miscellaneous	Clinical thermometer in metallic tube	1
	Water purification tablets (for 300–500 liters)	

The golden rule should be **use only what you know how to use**. Furthermore, in addition to observing the basic criteria for a modular first aid kit, those trekking without a doctor are urged to consider the needs of the native population.

Paragliding

Stéphane Oggier

Introduction

Paragliding (hang gliding with a parachute) (**252**), the newest airborne sport, has grown spectacularly in many Alpine countries. For example, there are over 20,000 paragliding enthusiasts in Switzerland, and some 50 schools. Because of the paraglider's weight (4–6 kg) and compactness, no more than a traditional backpack, the device is extremely simple to carry and can easily form part of a walker's equipment.

Until early 1988, the media were full of enthusiasm for paragliding, saying that it was an ingenious and inexpensive way to fly. However, the first accidents changed attitudes: during the summer of 1989, on average, 1–2 victims of serious accidents were admitted every day to Sion regional hospital, in the Western Alps.

252

252 The paraglider.

Background

Like the delta wing, the rectangular parachute is a product of the US Space Vehicle Research Agency, but it was in Europe, in the little French village of Mieussy, that the history of paragliding began.

In 1978 Jean-Claude Bétemps and Gérard Bosson had the idea of taking off from a mountain by running with a previously inflated canopy. In the following years, several world-famous mountain climbers discovered paragliding, contributing greatly to its fantastic surge in popularity.

Aerodynamics

If one had to define a paraglider from the aerodynamic point of view, one could say that it is a canopy that has several features in common with an aircraft frame (**253**).

The canopy flies because of its profile, kept in shape by air pressure entering its cells. The movement of the airflow around the profile creates low pressure at the canopy's upper surface and high pressure at the lower. These two associated phenomena create the lift that allows the paraglider to glide relatively well. The horizontal distance that can be covered by the most up-to-date canopies is around 5–6 m for every 1 m drop (= sensitivity of 6 m:1 m).

The major drawback of current paragliders is their lack of speed (less than 40 km/hour), which means they cannot be used in strong winds as they will be exposed to dangerous air conditions for longer periods. Their 'soft wing' design, inflated solely by the dynamic pressure of the air in which they are flying, is also a major weakness.

Incidence and causes of paragliding accidents

It is on take-off that 30% of paragliding accidents occur, many of which involve pilots in lower limb injuries. The most frequently observed causes are: falls during take-off; snagging immediately after take-off; return to the ground after settling on the seat (**254**); take-off in rotor turbulence (**255**) or with a following wind; snagged suspension lines; and contact with an obstacle.

253

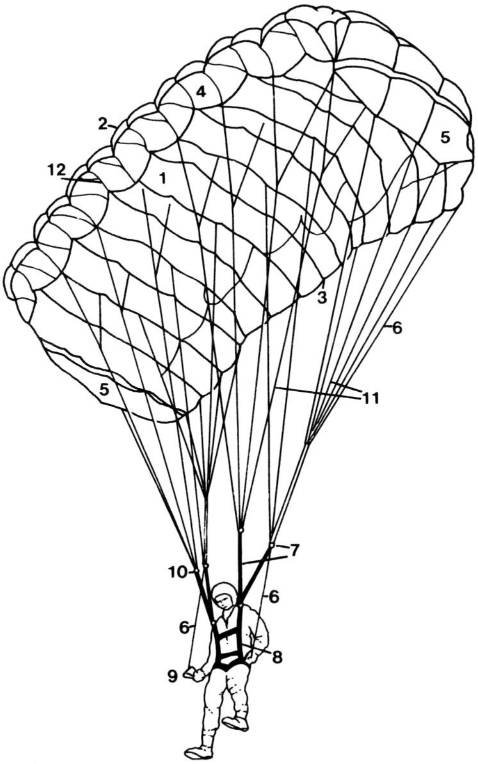

253 The structure of the paraglider. Canopy (1), Leading edge (2), Trailing edge (3), Cell (4), Stabilizer (5), Control line (6), Riser (7), Harness, cradle and seat (8), Control line handle (9), Carabiner and maillon connector (9), Suspension line (10), Cell walls (11).

254

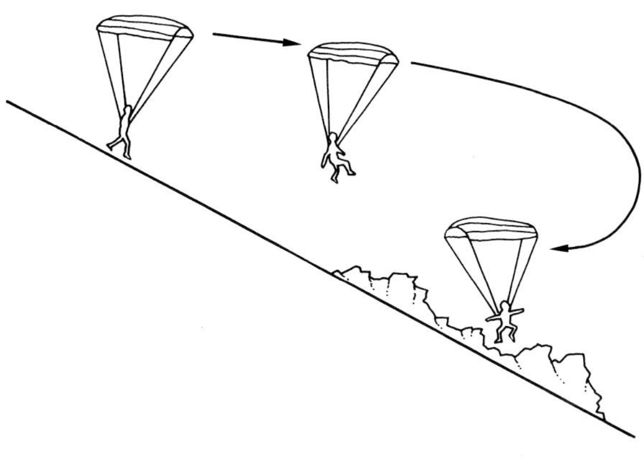

254 An accident upon take-off: the paraglider returns to the ground.

255

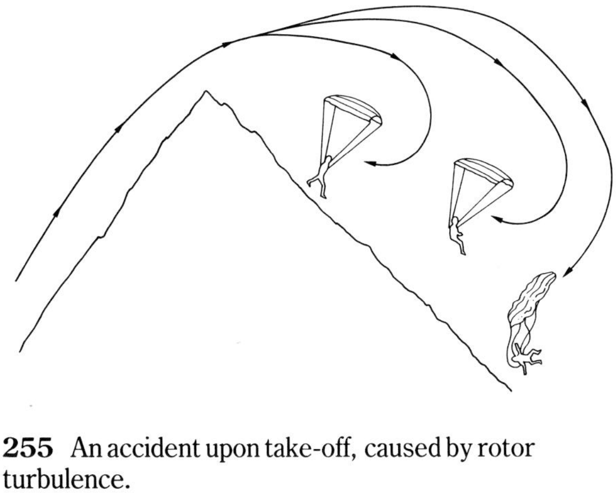

255 An accident upon take-off, caused by rotor turbulence.

25% of accidents occur during flight itself: rotors, curl-over turbulence under the wind (**256**), turbulence near outcrops, collision with obstacles (electrical wires, cables or another paraglider), and the canopy collapsing (together with autorotation), are the most frequent causes.

In competition the incidence of accidents after take-off rises to over 90% owing to the risks taken when looking for the slightest rising current and the high concentrations of paragliders in a given area. The percentage of accidents during flight also tends to rise among leisure pilots.

About 40% of accidents occur on landing, and result in a significant number of unstable spine lesions, with or without damage to the spinal cord. Upper and lower limbs are affected with equal frequency. These lesions generally occur as a result of a bad approach, which may arise from: landing in rotor turbulence (**257**); landing with a following wind or on a low turn; landing outside the landing area; collision with an obstacle; the canopy hooking or closing on approach (too much braking or rotor turbulence), and gradient wind speed (good wind on take-off, strong valley wind on landing) (**258**).

Finally, a small percentage of lesions (5%) occurs in other circumstances, such as cable-towed flights or ground trials in strong winds. Although the total number of accidents does seem to be stabilizing, that among active participants is rising sharply. In France, there were 5,000 participants in 1985, 15,000 in 1988 and over 20,000 in 1989. The current incidence of paragliding accidents has been estimated at around 3–4 accidents per 100 pilots annually, with fatalities at nearly 0.3 deaths per 1,000 annually. However, the number of deaths has decreased in comparison with previous years.

25

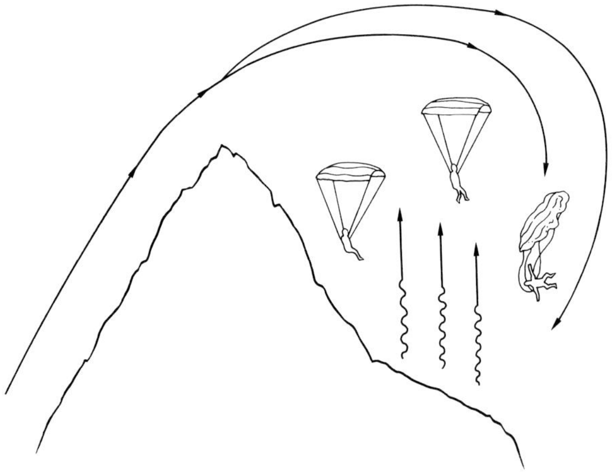

256 An accident during flight: the paraglider becomes unhooked, caught by thermal currents under the wing.

257

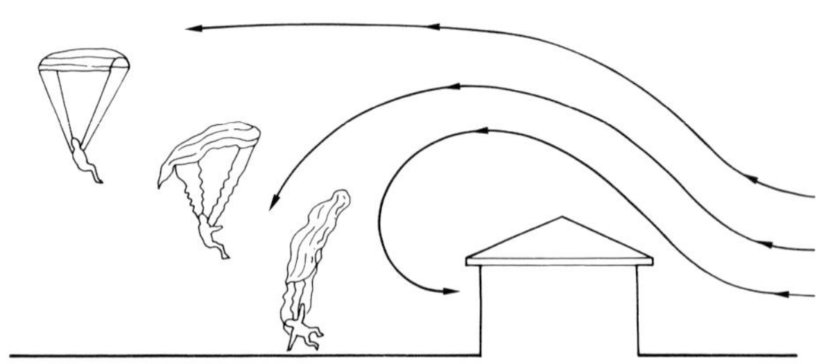

257 An accident upon landing: in the rotor turbulence caused by an obstacle, the paraglider becomes 'unhooked'.

258

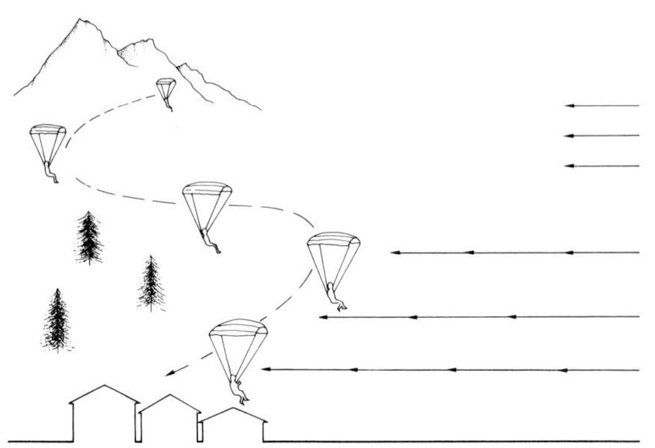

258 An accident upon landing: the pilot lands on some houses, caught unawares by a strong valley wind.

Summary of injuries

In a survey of lesions (**259**) carried out at Sion regional hospital in Switzerland, lower limb injuries predominated, accounting for more than 30% of the total (over 10% of patients also had an open fracture). The second most frequent lesion was spinal, accounting for more than 26% of lesions (38.2% of patients). The incidence of incomplete paraplegia is fairly high, but this type of paralysis is rarely reversible.

259

Order of frequency of main diagnoses (sometimes with associated lesions)

(Number of patients: 115)

Fractures of the spine column:	44
Head injuries:	15
Dorsal contusions:	13
Fractures of the ankle:	13
Sprains of the ankle:	12
Fractures of the tibia and/or fibula:	10
Fractures of the wrist:	9
Fractures of the pelvis:	8
Fractures of the femur:	7
Fractures of the radius and/or ulna:	7
Head or face wounds:	6
Fractures of the calcaneum:	4

Distribution of lesions
Number of lesions: 221

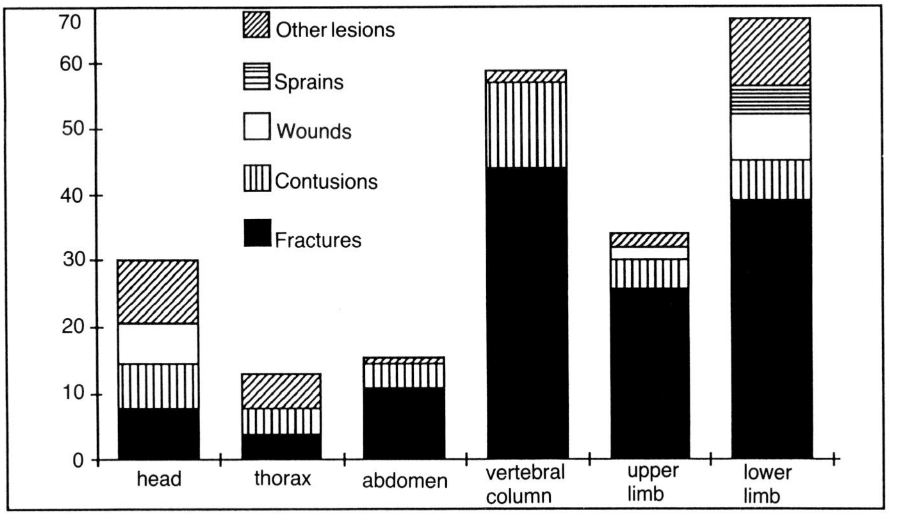

259 Main diagnoses of paragliding injuries (arranged in order of frequency) and the distribution of injuries.

Risk factors

Most paragliding accidents are caused by the pilots themselves, owing to poor assessment of the situation, major technical faults, and inadequate theoretical and practical preparation. However, accidents are also brought about by factors outside the pilots' control, such as the type of canopy, the terrain and the season.

The number of completed jumps seems to be the key element in a pilot's experience (**260**). Thus, more than two out of three accidents occur during the initial training period, covering the first 100 jumps.

A second peak in the frequency of accidents has been noted at around 200 flights, when acquired experience and self-confidence spur the pilot on to explore more difficult meteorological and geographical conditions. Accidents among experienced pilots who have completed over 500 flights are related to their search for very high performance.

The pilot's licence is obtained after a theoretical examination and a demanding practical test, but it does not constitute an infallible guarantee against accidents. However, it should be noted that studies have demonstrated a very high incidence of serious injuries in pilots operating without a licence.

Very few accidents can be blamed on faulty equipment. Good equipment is certainly currently available, but the standard of the pilots is sometimes questionable. Potentially dangerous devices are those that are deliberately chosen to be oversized in relation to the pilot's weight, or those that are too technically advanced for his or her skills. Qualified pilots with high-performance canopies do not seem to be more seriously or more frequently injured than others. However, it is possible that certain types of seat harness can indirectly aggravate dorsolumbar trauma: during difficult landings, several people have been known to stay on their seats instead of going into a vertical position. The shock on landing is therefore transmitted through the axis of the spine. These harnesses could therefore be greatly improved in terms of the protection they afford against dorsolumbar trauma.

The meteorological situation, season, and time of day when accidents occur, are key risk factors: poor assessment of the weather conditions, for example failing to spot the approach of atmospheric disturbance, has had very serious consequences, even with experienced pilots. In Switzerland, it has been observed that around 100 days in the year are favorable for paragliding, 100 are clearly unfavorable, and during the rest of the year, weather conditions are doubtful. It is during this last period that a significant number of accidents take place.

Winter seems to be the best season for 'high-safety' paragliding: the cold and dense air creates less turbulence, and the snow is a good cushion for falls, providing take-off and landing sites that are smooth and very reliable. If you are not a pilot, but would like to become one, start learning in the winter over snow.

260

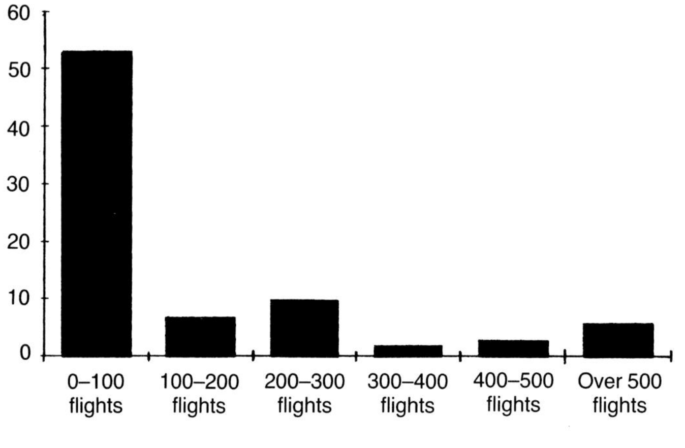

260 Number of accidents in relation to number of completed flights among 81 pilots.

Prevention

There are three types of advice that can be given to paragliding pilots:

- **The pilot himself:** never fly without a licence; systematically carry out a proper check before leaving; never leave if it is not possible to abandon take-off, e.g. in high winds; always take suitable footwear, a helmet and a walkie-talkie; never fly alone or without supervision; learn the flight priority rules, and read and learn as much as you can about meteorology.
- **The weather:** always get weather information before a flight; beware of valley wind and rotors under the wind; keep away from outcrops and other aircraft, and never fly in very unsettled weather.
- **Equipment:** never fly with a canopy that is unsuitable or too advanced for you to operate without supervision, and never modify a canopy without expert advice.

A golden rule could be: **know when to give up**. True courage is displayed not by taking-off at any cost, but by knowing when to fold up your canopy calmly if you feel that a flight would involve risk (even after three hours' walk to get to the take-off site).

Conclusion

The ever-improving performance of paragliders now makes it possible to use ascending currents effectively, with minimal physical effort and great pleasure. The great danger of paragliding nowadays is that it looks so simple and appears so safe. Consequently, many pilots have had serious accidents, even when convinced that they were being prudent. Sound theoretical and practical training, with regular refresher courses, is absolutely essential.

Looking at current statistics, it is possible to be either pessimistic or optimistic. One can be pessimistic when thinking about the thousands of paragliding pilots with the minimum level of training, who are going to find themselves in springtime turbulence and not know what to do. However, when one realizes that most serious accidents are often due to stupidity, one can be more optimistic: by training pilots, improving schools and equipment and, at the same time encouraging people to take a guide with them as in mountain climbing, the number of accidents will decrease. There are no risks that cannot be diminished, and if you do not want to fly, there is nothing to stop you dreaming!

Future prospects of mountain medicine

Frédéric Dubas and Jacques Vallotton

The generation of explorer–climbers of the last century and the beginning of this century has now been succeeded by a generation of athletic mountaineers who are spurred on not only by a taste for adventure but, more frequently, by a spirit of competition that is probably connected to the ever-growing importance of the media and sponsorship. Skiers and climbers are now performing at the limits of human endurance.

With the advent of a more leisure-orientated life style, mountain sports have become more accessible to everyone and new sports have made their appearance, such as competition rock climbing, free rock climbing, monoskiing, snowboarding, paragliding and abseiling, to mention only the best known. While the equipment has become more sophisticated, resulting in greater safety, the increased risks involved in certain sports explain the growing incidence and seriousness of accidents and the high energy injuries that are occurring, particularly among skiers.

Lack of preparation, a misunderstanding of one's own abilities and failure to recognise the objective dangers of the surroundings are to blame for the rise in accidents in the mountains, a large number of which could be avoided if certain basic rules were observed. It is, therefore, extremely important to develop preventive measures and to promote wide-ranging information campaigns aimed at climbers and skiers. Various associations have made it their objective to publicize certain guidelines, aimed at improving safety in the mountains. Among these, the International Society for Mountain Medicine (ISMM), the medical commission of the International Union of Alpinists' Associations (UIAA) and the mountain rescue subcommittee of the International Commission for Mountain Rescue (CISA) should be mentioned.

The growing number and seriousness of accidents in the mountains is, in turn, increasing the risks to the rescuers, necessitating a top quality organisation and improved resources. Mountain rescue has therefore become a specific field, and in the same way, because of the skills it demands, mountain medicine is on the way to becoming a special branch of medicine.

In the future, the aim of mountain medicine must be to improve a patient's chances of survival in hostile surroundings, and to rescue the increasing number of accident victims in the most effective way. To this end, efforts in all fields of research must be coordinated and international cooperation between existing organisations is of paramount importance.

Rescue operations must be rationally organized and adapted to the territory and to the structures available. The success of an operation depends on certain key factors, namely the speed with which the alarm is raised, the clarity of the messages that are transmitted, the speed of response, the availability and skill of the rescuers, and their knowledge of the terrain. Certain technological innovations are now improving the chances of success, and in the next few years we can expect to see many more advances in all fields of mountain medicine.

Current developments include:

• **Detectors of victims of avalanches (DVA)** are transceivers of reduced size and weight, allowing rapid location of the victim of an avalanche (provided, of course, that the victim is also carrying a transceiver). It is to be hoped that the various devices available on the market, e.g. Autophon VS-68, Orthovox, Pieps, Redar and Arva (**261**), will all work on the same wavelength in future and have the same technical features. Their price should drop to allow them to come into general use, and there should be wide-scale instruction on how to operate them.

• **Radio transmitters only** (ARVA (Appareil de Recherche de Victimes d'Avalanches) belt) or **reflectors by means of a diode** (RECCO system). While the use of such devices may be accepted unreservedly in certain supervized areas of skiing, where the rescuers can be at the scene of the accident within a few minutes, their use off the beaten track cannot be recommended, since it would necessitate the involvement of a rescue organization that had been supplied with the corresponding detection devices. Under such conditions, these devices can give a false feeling of security and have no advantage over avalanche dogs.

• **The thermal parachute** (Société Gabriel, 65, Rue du Souvenir, F-69261 Lyons) was developed from an original idea by Dr Jacques Foray of Chamonix. It helps to reduce, and even to prevent, heat loss when a victim is subjected to conditions of intense cold, especially in cases of falls into crevasses (**262**).

• **The portable hyperbaric bag** (Certec SA, F-69210, Sourcieux Les Mines) (**263**) enables victims of mountain sickness to experience simulated atmospheric conditions equivalent to those at lower altitudes, as part of their treatment.

261

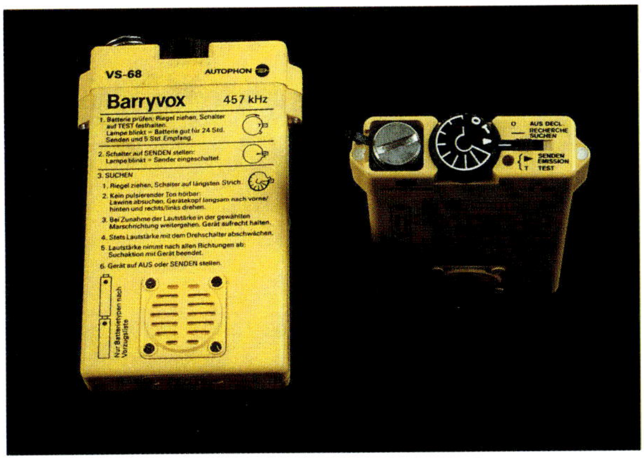

261 The Autophon VS-68 device, also called Barryvox, is one of the standard avalanche victim detection devices.

262

262 The thermal parachute transports warmed, humidified air to the trapped victim of a fall into a crevasse. *Courtesy of J. Foray.*

263 

263 The portable hyperbaric bag is actually the best treatment for acute mountain sickness when a rapid descent is impossible.

While the helicopter has revolutionized mountain rescues, the medical part of the operation is now of the utmost importance, as has been proven in road accidents. The constant increase in multiple injuries, not only in climbing accidents, but also in the case of skiers and abseilers, justifies more and more the presence of a doctor alongside the rescuers. Certain devices that are currently available facilitate a more precise diagnosis of injuries, and aid the decision as to the most appropriate treatment:

- **The Oxypuls** (Ohmeda, 1315 West Century Drive, Louisville, CO 80027, USA) enables an immediate assessment of the pulse and of the capillary saturation in oxygen (**264**).
- **The Oxylog** (Draeger–Werke AG, P.O. Box 1339, Moislingerallee, 53–55, D-2400 Lübeck) is a portable ventilator that enables a gaseous mixture (air–oxygen or pure oxygen) to be administered at a constant pressure from a tank or another source (**265**). The mixture, the frequency and the volume that is administered can be adjusted. The device is very simple to operate, making it highly suitable for use during rescues.

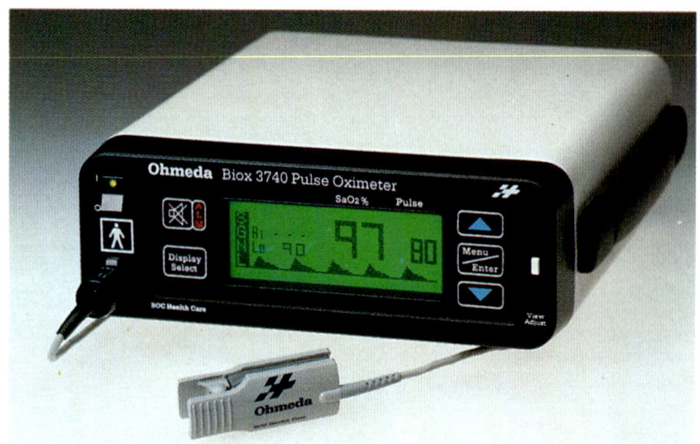 264

264 The Ohmeda–Oxypuls device.

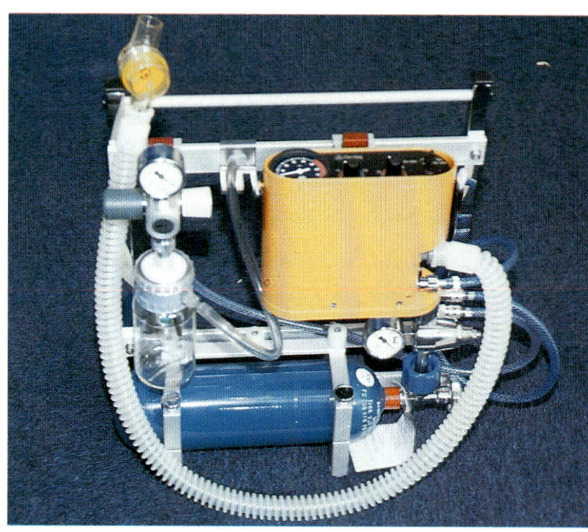 265

265 The Draeger–Oxylog device.

- **The Schiller miniscope MS-2** (Schiller AG, 68, Altgasse, CH-6340 Baar) enables the heart rate to be assessed by simple contact with the wall of the chest: its small size and reduced weight guarantee it a place in the action packs (**266**).
- **The epitympanic thermometer** (Georges Mettraux, 1023-Crissier, Switzerland), developed from an idea by Professor Lomax and Dr Dubas, has made the measurement of core temperature easy (**267**).

The thermometer, which is fitted into a head band and equipped with a flexible probe that is placed almost in contact with the ear drum, provides a good estimate of the temperature of the brain within a few minutes. Such a measurement, taken at the scene of the accident, allows better assessment of the state of the victim and reduces delays in getting him or her to a suitable place of treatment.

266

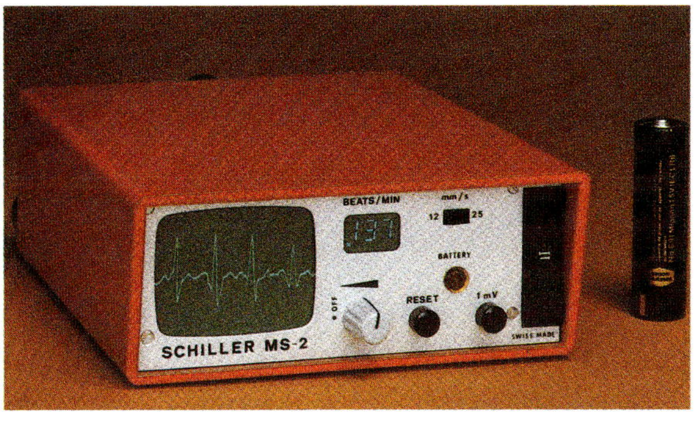

267

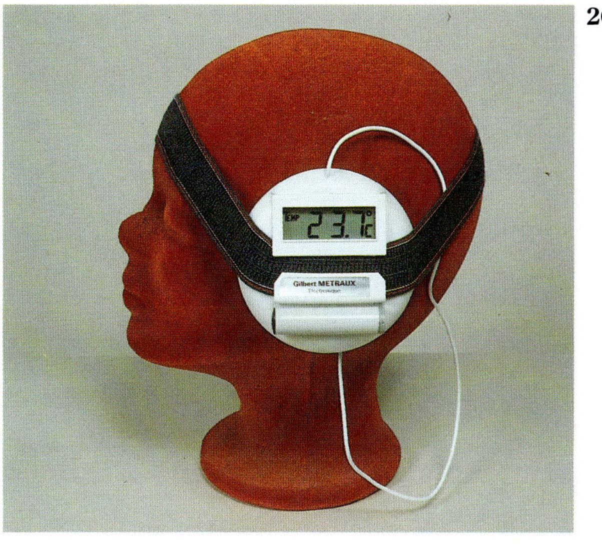

266 The Schiller miniscope monitoring device.

267 The Dubas–Mettraux epitympanic thermometer.

Avalanches pose problems for rescuers that are still far from being solved. Could we possibly imagine a scanner that would enable a mass of snow to be rapidly explored in a search for a buried victim?

Although some questions remain, we may expect, in the future, to find the answers and to obtain a consensus over certain points that are currently controversial.

Nevertheless, the authors are convinced that only preventive measures can reduce the increasing number of deaths and accidents in the mountains. Stress must be laid on the importance of: informing the public (e.g. weather and avalanche forecasts); the preventive release of avalanches; training of the rescue teams; marking and maintenance of the ski slopes; and of any measures that continue to improve the safety of hikers.

References

Early scientific expeditions to high altitude

Barcroft J, Binger CA, Bock AV, *et al.* Observations upon the effect of high altitude on the physiological processes of the human body, carried out in the Peruvian Andes, chiefly at Cerro de Pasco. *Phil Trans R Soc Ser B* 1923; **211**: 351–480.

Dill DB, Talbott JH, and Consolazio WV. Blood as a physiocochemical system. *J Biol Chem* 1937; **118**: 649–666.

Douglas CG, Haldane JA, Henderson Y, and Schneider EC. Physiological observations made on Pike's Peak, Colorado, with special reference to adaptation to low barometric pressures. *Phil Trans R. Soc Ser B* 1913; **203**: 185–318.

Filippi FD. *Karakorum and Western Himalaya.* London: Constable, 1912.

Freshfield DW. *The Life of Horace Benedict de Sassure.* London: Edward Arnold, 1920.

Houston CS, Sutton JR, Cymerman A, and Reeves JT. Operation Everest II: man at extreme altitude. *J Appl Physiol* 1987; **63**: 877–882.

Pugh LGCE, Gill MB, Lahiri S, Milledge JS, Ward MP, and West JB. Muscular exercise at great altitude. *J Appl Physiol* 1964; **19**: 431–440.

West JB. *Everest – The Testing Place.* New York: McGraw-Hill, 1985.

West JB. Highest inhabitants in the world. *Nature* 1986; **324**: 517.

West JB and Alexander M. Kellas and the physiological challenge of Mount Everest. *J Appl Physiol* 1987; **63**: 3–11.

Mountain rescue: modern strategies

Durrer B. Winch rescues. Swiss Air Rescue (REGA) 1983–1984. *Proceedings of the International Aeromedical Evacuation Congress* 1985.

Yearbooks. Swiss Air Rescue. 1980–1990.

Definition and parameters

Brunt D. *Physical and Dynamical Meteorology* 2nd edn. Cambridge: Cambridge University Press, 1952, 379.

Cerretelli P. Gas exchange at high altitude. In: West JB, ed. *Pulmonary Gas Exchange* (Vol. 2). New York: Academic Press, 1980; pp. 97–147.

Cruz J. Mechanics of breathing in high altitude and sea level subjects. *Respir Physiol* 1973; **17**: 146–161.

Dill DB, and Evans DS. Report barometric pressure! *J Appl Physiol* 1970; **29**: 914–916.

Houston CS. Going High. In: *The story of man and altitude.* New York: American Alpine Club, 1980.

Pugh LGCE. Resting ventilation and alveolar air on Mount Everest: with remarks on the relation of barometric pressure to altitude in mountains. *J Physiol* 1957; **135**: 590–610.

Varène P, Timbal J, and Jacquemin C. Effect of different ambient pressures on airway resistance. *J Appl Physiol* 1967; **22**: 699–706.

West JB, Lahiri S, Maret KH, Peters Jr RM, and Pizzo C. Barometric pressure at extreme altitudes on Mount Everest: Physiological significance. *J Appl Physiol* 1983; **54**: 1188–1194.

Nutrition at altitude

Boyer SJ, and Blume FD. Weight loss and changes in body composition at high altitude. *J Appl Physiol* 1984; **57**: 1580–1585.

Consolazio CF, Matoush LO, Johnson HL, Krzywicki HJ, and Daws TA. Effects of high carbohydrate diets on performance and clinical symptomatology after rapid ascent to high altitude. *Fed Proc* 1969; **28**: 937–943.

Hoppeler H, Kleinert E, Schlegel C, Claassen H, Howald H, Kayar SR, and Cerretelli P. II. Morphological adaptations of human skeletal muscle to chronic hypoxia. *Int J Sports Med* 1990; **11** (suppl 1) 53–59.

Sleep at altitude

Mosso A. *Life of man in the high Alps.* London: T Fisher Unwin, 1898.

Natani K, Shurley JT, Pierce CM, and Brooks RE. Long-term changes in sleep patterns in man on the South Polar plateau. *Arch Intern Med* 1970; **125**: 655–659.

Powles ACP. Sleep at altitude. In: Sutton JR, Jones NL, and Houston CS, eds. *Hypoxia: Man at altitude.* New York: Thieme Verlag, 1982; pp. 182–185.

Sutton JR, Houston CS, Mansell AL, McFadden MD, Hackett PM, Rigg JRA, and Powles ACP. Effect of acetazolamide on hypoxemia during sleep at high altitude. *N Engl J Med* 1979; **310**: 1329–1331.

West JB, Peters Jr RM, Aksnes G, Maret KH, Milledge JS, and Schoene RS. Nocturnal periodic breathing at altitudes of 6300 and 8050 m. *J Appl Physiol* 1986; **61**: 280–287.

Acute mountain sickness

Ferrazzini G, Maggiorini M, Kriemler S, Bärtsch P, and Oelz O. Successful treatment of acute mountain sickness with dexamethasone. *BMJ* 1987; **294**: 1380–1382.

Hackett PH, Rennie D, and Levine HD. The incidence, importance, and prophylaxis of acute mountain sickness. *Lancet* 1976; **2**: 1149–1154.

Hackett PH. *Mountain sickness. Prevention, recognition and treatment.* New York: American Alpine Club, 1980.

Hackett PH and Roach RC. Medical therapy of altitude illness. *Ann Emerg Med* 1987; **16**: 980–986.

Johnson TS and Rock PB. Acute mountain sickness. *N Engl J Med* 1988; **319**: 841–845.

Acute mountain sickness: risk factors

Hu S-T, Huang S-Y, Chu S-C, and Pa CF. Chemoreflexive ventilatory response at sea level in subjects with past history of good acclimatization and severe acute mountain sickness. In: Brendel W and Zinc RA, eds. *High-Altitude Physiology and Medicine.* New York, Heidelberg, Berlin: Springer-Verlag, 1982; pp. 28–32.

King AB and Robinson SM. Ventilation response to hypoxia and acute mountain sickness. *Aerospace Med* 1972; **43**(4): 410–421.

Masuyama S, Kimura H, Sugita T, *et al.* Control of ventilation in extreme-altitude climbers. *J Appl Physiol* 1986; **61**(2): 500–506.

Mathew L, Gopinathan PM, Purkayastha SS, Sen Gupta J, and Nayar HS. Chemoreceptor sensitivity and maladaptation to high altitude in man. *Eur J Appl Physiol* 1983; **51**: 137–144.

Moore LG, Harrison GL, McCullough RE, *et al.* Low acute hypoxic ventilatory response and hypoxic depression in acute altitude sickness. *J Appl Physiol* 1986; **60**(4): 1407–1412.

Oelz O, Howald H, Di Prampero PE, *et al.* Physiological profile of world-class high-altitude climbers. *J Appl Physiol* 1986; **60**(5): 1734–1742.

Richalet J-P, Corizzi F, Kéromès A, *et al.* Maximal O_2 consumption as a determinant of performance in extreme altitude. In: Sutton JR, Houston CS, and Coates G, eds. *Hypoxia. The tolerable limits.* Indianapolis: Benchmark Press, 1988.

Richalet J-P, Kéromès A, Dersch B, *et al.* Caractéristiques physiologiques des alpinistes de haute altitude. *Science et Sports* 1988; **3**: 89–108.

Schoene RB, Lahiri S, Hackett PH, *et al.* Relationship of hypoxic ventilatory response to exercise performance on Mount Everest. *J Appl Physiol* 1984; **56**(6): 1478–1483.

Viswanathan R, Subramanian S, Lodi STK, and Radha TG. Further studies on pulmonary oedema of high altitude. *Respiration* 1978; **35**: 216–222.

Vizek M, Pickett CK, and Weil JV. Interindividual variation in hypoxic ventilatory response: potential role of carotid body. *J Appl Physiol* 1987; **63**(5): 1884–1889.

Localized or peripheral edema

Hackett PH and Rennie D. Rales, peripheral edema, retinal hemorrhage and acute mountain sickness. *Am J Med* 1979; **67**: 214–218.

Maggiorini M, Bühler B, Walter M, and Oelz O. Prevalence of acute mountain sickness in the Swiss Alps. *BMJ* 1990; **301**: 853–855.

Polycythemia and thromboembolic episodes

Dickinson JD, Heath D, Gosney J, and Williams D. Altitude-related deaths in seven trekkers in the Himalayas. *Thorax* 1983; **38**: 646–656.

High-altitude pulmonary edema

Hackett PH. *Mountain sickness. Prevention, recognition and treatment.* New York: American Alpine Club, 1980.

Oelz O, Maggiorini M, Ritter M, Waber U, Jenni R, Vock P, and Bärtsch P. Nifepidine for high altitude pulmonary oedema. *Lancet* 1989; **2**: 1241–1244.

Schoene RB. Pulmonary edema at high altitude. Review, patho-physiology, and update. *Clin Chest Med* 1985; **6**: 491–507.

Mental and neurological disturbances at high altitude

Barragán ME, Arce J. Influencia de la altura en el sistema nervioso central (SNC). In: *Instituto Boliviano de Biologia de Altura. Bodas de Plata 1963–1988.* Facultad de Medicina UMSA, La Paz, 1988: 142–173.

Clarke C. High-altitude cerebral oedema. *Int J Sports Med* 1988; **9**(2): 170–174.

Hajdukiewicz J and Ryn Z. Höhenhirnödem. Medizinische Probleme bei Bergfahrten in grossere höhen. *Internationale Bergrettungs-Aerzte-Tagung.* Innsbruck 15 Nov 1980. Innsbruck: G. Flora, 1983; 88–92.

Monge CM and Pesce H. El sistema nervioso vegetativo del hombre de los Andes. *Anales de la Facultad de Ciencias Médicas.* UNMSM, Lima, 1935; **17–18**: 43–59.

Querol M. The electroencephalogram in a group of native highlanders at 4540 m altitude and at sea level. *Electroenceph Clin Neurophysiol,* 1965; **18**(4): 401–408.

Ramos A, Kruger H, Maro M, and Arias-Stella J. Patología del hombre nativo de las grandes alturas – investigación de las causas de muerte en 300 autopsias. *Boletin Oficina Sanitaria Panamericana* 1967; **62**: 496.

Ryn Z. High altitude and brain damage. *Abstracts from UIAA Mountain Medicine Conference* 19–20 Nov 1987, St. Bartholomew's Hospital, London.

Ryn Z. High altitude cerebral asthenia. *Int J Sports Med* 1989.

Ryn Z. Nervous system and altitude. Syndrome of high altitude asthenia. *Acta Med Polona* 1979; **20**(2): 155–169.

Severinghaus JW. The role of the cerebrospinal fluid in human acclimatization to high altitude. *Ann Intern Med* 1963; **58**(4): 729–732.

Velázquez T. *Correlation between altitude and consciousness time in high-altitude natives.* Texas: Brooks Air Force Base, USAF School of Aviation Medicine, 1959.

The effects of solar radiation

Duke-Elder WS, and MacFaul PA. Radiational injuries. In: Duke-Elder WS, ed. *System of Ophthalmology* (Vol. 14, Part 2) St. Louis: CV Mosby Co, 1972; pp. 918–928.

Pitts DG. A comparative study of the effects of ultraviolet radiation on the eye. *Am J Optom* 1970; **47**: 535.

Pitts DG, Cullen AP, and Hacker PD. Ocular effects of ultraviolet radiation from 295 to 365 nm. *Invest. Ophthalmol. Visual Sci* 1977; **16**: 932.

Pitts DG and Tredici TJ. The effects of ultraviolet light on the eye. *Am Ind Hyg Assoc J* 1977; **32**: 235.

Records RC and Brown JL. Light and photometry. In: Duane TE and Jaeger E, eds. *Biomedical Foundations of Ophthalmology*. Philadelphia: Harper and Row, 1986 (Vol. 2).

Scotto J, Frears TR, and Fraumeni JF. Solar radiation. In: Schottenfeld D and Fraumoni JF, eds. *Cancer Epidemiology and Prevention*. Philadelphia: WB Saunders Co, 1982; pp. 254–276.

Frostbite

Ariev IJ. *Monograph on Frostbite (1940)*. Canada: Defence Research Board, [Translation, 1955]: 171.

Campbell MR. *Proceedings, Frostbite Symposium*. Arctic Aero Med Lab, Ft Wainwright, Alaska, Feb, 1964.

Foray J, Binder F, and Alouso JP. Les gelures de montagnes. *Chirurgie* 1977; **103**: 98–109.

Franz DR, Berberich JJ, Blake S, and Mills WJ. Evaluation of fasciotomy and vasodilator for treatment of frostbite in the dog. *Cryobiology* 1978; **15**: 659–669.

Kappes BA and Mills WJ. Thermal biofeedback in the treatment of frostbite and cold injuries. *Am J Clin Biofeedback* 1982; **5**(1).

Karow AM and Watts RW. Tissue freezing. *Cryobiology* 1965; **2**(3).

Killian H. *Cold Injuries with Special Reference to German Experience in WWII*. Edition Cantor KG–Aulendorf & Wurtt, [Translation], 1952.

King RD, Kaiser GC, Lempke RE, and Schumacker Jr HB. Evaluation of lumbar sympathetic denervation. *Arch Surg* 1964; **88**: 36.

Lange K, Boyd L, and Loewe L. Functional pathology of frostbite and prevention of gangrene in experimental animals and humans. *Science* 1945; **102**(151).

MaCauley RB, Hing D, Robson M, and Heggers JP. Frostbite injuries: A rational approach based on pathophysiology. *J Trauma* 1983; **22**(2).

Meryman HT. Tissue freezing and local cold injury. *Physio Rev* 1957; **37**(2): 233–252.

Meryman HT. *Cryobiology*. Academic Press, 1966: pp. 48–88; 90–102.

Mills WJ. Frostbite. *Alaska Medicine* 1983; **25**(2): 33–38.

Mills WJ, *et al*. Medical aspects of mountain climbing, Round Table, *The Physician and Sports Medicine* 1977; **5**(3).

Mills WJ, Whaley R, and Fish W. Frostbite, experience with rapid rewarming and ultrasonic therapy. *Alaska Medicine* 1960; **2**(1), **2**(4) 1961; **3**(2).

Quintanella RF, Krusen H, and Essex HE. Studies on frostbite with special reference to treatment and the effect on minute blood vessels. *Am J Physiol* 1947; **149**: 149.

Sutton JR, Houston CS, and Coates G, eds. *Hypoxia and Cold*. New York: Praeger Publishing, 1987; pp. 340–362.

Hypothermia

Danzl DF and Pozos RF. Multicenter hypothermia survey. *Ann Emerg Med* 1987; **16**: 1042–1055.

Golden FStC and Hervey GR. The 'after-drop' and death after rescue from immersion in cold water. In: Adams JN, ed. *Hypothermia Ashore and Afloat*. Aberdeen: Aberdeen University Press, 1981: pp. 37–56.

Harari H, Regnier B, Rapin M, Lemaire F, and Le Gall JR. Hemodynamic study of prolonged deep accidental hypothermia. *Eur J Intensive Care Med* 1975; **1**: 65–70.

Hauty MG, Esrig BC, Hill JG, and Long WB. Prognostic factors in severe accidental hypothermia: experience from the Mount Hood tragedy. *J Trauma* 1987; **27**: 1107–1112.

Jessen K and Hagelsten JO. Peritoneal dialysis in the treatment of profound accidental hypothermia. *Aviat Space Environ Med* 1978; **49**: 426–429.

Jurkovich GJ, Greiser WB, Luterman A, and Curreri PW. Hypothermia in trauma victims: an ominous predictor of survival. *J Trauma* 1987; **27**: 1019–1024.

Ledingham IMcA and Mone JG. Treatment of accidental hypothermia: a prospective clinical study. *BMJ* 1980; **1**: 1102–1105.

Lilly JK, Boland JB, and Zekan S. Urinary bladder temperature monitoring: a new index of body core temperature. *Crit Care Med* 1980; **8**: 742–744.

Lloyd EL. *Hypothermia and Cold Stress.* Rockville, Maryland: Aspen Publications, 1986.

Paton BC. Accidental hypothermia. *Pharmac Ther* 1983; **22**: 331–337.

Pozos RS, Wittmers LE, eds. *The nature and treatment of hypothermia.* Minneapolis: University of Minnesota Press, 1983.

Splittberger FH, Talbert JG, Sweezer WP, and Wilson RF. Partial cardiopulmonary bypass for core rewarming in profound accidental hypothermia. *Am Surg* 1986; **52**: 407–412.

White FN. A comparative physiological approach to hypothermia. *J Thorac Cardiovasc Surg* 1981; **82**: 821–831.

Zell SC and Kurtz KJ. Severe exposure hypothermia: a resuscitation protocol. *Ann Emerg Med* 1985; **14**: 428–431.

The management of skiing accidents

Bèzes H and Massart P. Les lésions du membre supérieur par accident de ski. 2,930 observations sur 13,250 accidents. *J Traumatol Sport* 1985; **2**: 17–27.

Bèzes H and Massart P. Les accidents du ski de fond. D'après 753 observations. *J Traumatol Sport* 1986; **3**: 70–85.

Bèzes H, Massart P, and Fourquet J-P. Die Osteosynthese der Calcaneus-Impressionsfraktur. Indikation, Technik un Resultate bei 120 Fällen. *Unfall-Heilkunde* 1984; **87**: 363–368.

Guyot C. 20 ans d'accidents de ski à l'Hôpital-Sud de Grenoble (d'après une statistique de 17,120 accidentés 1968–1988) [Thesis]. Grenoble: Faculté de Médécine, Université de Grenoble, 1989.

Massart P, and Bèzes H. L'entorse grave métacarpo-phalangienne du pouce au cours des accidents de ski. A propos de 125 réparations chirurgicales sur un ensemble de 340 entorses métacarpo-phalangiennes du pouce par accident de ski. *Ann Chir Main* 1984; **3**: 101–112.

Capsuloligamentary injuries to the metacarpophalangeal joint of the thumb ('skier's thumb')

Bowers WH, and Hurst LC. Gamekeeper's thumb. Evaluation by arthrography and stress roentgenography. *J Bone Jt Surg* 1977; **59A**: 519–524.

Cantero J, Cruz M, and Perrenoud L. Les lésions capsuloligamentaires récentes de l'articulation métacarpophalangienne du pouce. *Ann Chir* 1980; **34**(9): 655–662.

Glas K. Ski-Daumen. Morphologische Grundlagen zur Bänderverletzung im Bereich des Metacarpo-Phalangeal Gelenkes am Daumen. *Fortschr Med* 1978; **96**(4): 185–187.

Kapandji IA. *Physiologie articulaire.* Fascicule II. Librairie Maloine, éd. Paris, 1966.

Stener B. Displacement of the ruptured ulnar collateral ligament of the metacarpophalangeal joint of the thumb. A clinical and anatomical study. *J Bone Jt Surg* 1962; **44B**: 869–879.

Verdan C. *Traitement chirurgical des séquelles des entorses métacarpophalangiennes du pouce. Traumatismes ostéoarticulaires de la main.* GEM. L'Expansion Scientifique Française, éd. Paris, 1971: 115–121.

Monoskiing and surfing: new risks

Binet MH, Ryvlin P, and Fairbrother R. Nouvelles glisses, nouveaux risques. *Seventh ISSS Symposium,* Kiruna, to be published by ASTM.

Montillet B, Binet MH, and Delouche G. Les accidents de monoski alpin. *J de Traumatologie du Sport* 1987; **4**: 72–76.

Shealy D. Surf accidents. *Sixth ISSS Symposium,* Naeba, Japan, to be published by ASTM.

Forces on the human body during falls on the rope and their consequences

Burton RR. Operational G-induced loss of consciousness: something old, something new. *Aviat Space Environ Med* Aug 1985, Aerospace Med Ass, Washington DC.

Henzel JF. *The human spinal column and upward ejection acceleration: An appraisal of biodynamic implications.* Ohio: Aerospace Medical Research Laboratories, 1967.

Mägdefrau H. Die Hüftanseilmethode und ihre Gefahren. *Praktische Sport-traumatologie und Sportmedizin.* Munich: Zuckschwerdt, 2/1987.

Mägdefrau H. *Die Belastung des menschlichen Körpers beim Sturz ins Seil und deren Folgen* [Dissertation]. Munich, 1990.

Martinez Villen G. Can harnesses be dangerous to the climber? Conference paper, Zaragoza, 1987.

Schubert P. *Die Hüftanseilmethode.* Munich: DAV-Mitteilungen, 1984.

Webb P. *Bioastronautics Data Book.* NASA-SP-3006, 1964.

Injuries in sport climbing

Bollen SR. Injury to the A2 pulley in rock climbers. *J Hand Surg* (British Volume) 1990; **15B**: 268–270.

Bollen SR. Soft tissue injury in extreme rock climbers. *Br J Sports Med* 1988; **22**(4): 145–147.

Bowie WS, Hunt TK, and Allen Jr HA. Rock climbing injuries in Yosemite National Park. *West J Med* 1988; **149** (Aug): 172–177.

Brniak M. Speed climbing. *Climber & Rambler* 1979; **18**(3) (Mar): 22–24.

Bunting CJ, *et al.* Stress in the adventure activities of rock climbing and vappelling. *J Sports Med Physical Fitness* 1986; **26**(1): 15–20.

Burtscher M, and Jenny E. *Most common complaints and injuries in free-climbers conditioned by training.* UIAA International Symposium, Munich, Oct 1986.

Cole AT. Fingertip injuries in rock climbers. *Br J Sports Med* 1990; **24**(1): 14.

Darlot P. *Pathologie de la main liée à la pratique de l'escalade* [Thesis]. Paris: Faculté de Médécine, Université Pierre et Marie Curie, 1986.

Della Santa DR and Kunz A. Le syndrome de surchage digitale lié à l'escalade sportive. *Schweiz Z Sportmed* 1990; **38**(1): 8–9.

Dobyus JH. Sport stress syndromes of the hand and wrist. *Am J Sports Med* 1978; **6**(5): 236–254.

Donnelly P. Subcultural reproduction and transformation in climbing. *International Review for the Sociology of Sport* 1985; **20**(2): 3–15.

Donnelly P. Social climbing: a case study of the changing class structure of rock climbing and mountaineering in Britain. *Second North American Society for the Sociology of Sport Conference.* Fort Worth, Texas, 1981.

Horterer H and Hipp E. Chondropathy of the patella caused by mountain climbing. *Fortschr Med* 1983; **101**(5) (Feb): 157–159.

Leal C, Rañé A, and Herrero R. Soziologie, Trainingszeit und Fingerverlungen beim Sportkletteren. *Praktische Sport-Traumatologie und Sportmedizin* 1987; **2**: 44–47.

Lin GT, Amadio KN, Cooney WP, and Cheo EYS. Biomechanical analysis of finger pulley reconstruction. *J Hand Surg* 1989; **14B**(3).

Lin GT, Cooney WP, Amadio PC, and An K. *J Hand Surg* 1990; **15B**(4).

McKeag DB. The concept of overuse. The primary care aspects of overuse syndromes in sports. *Primary Care* 1984; **11**(1): 43–59.

Radlinger L. Entrainement sportif: Conseils et instructions à l'intention des grimpeurs. *Les Alpes* 1984; **12** (Dec): 517–519.

Reynard E. Aspects pathologiques du membre superieur lors de la pratique de l'escalade [Thesis]. Nice: Faculté de Médecine, Université de Nice, 1985.

Rotman I. *Overuse injuries of the hand in sport climbing.* Information Sheet of the Medical Commission of the UIAA, 1986.

Rotman I. Overuse syndromes of the hand in top Czechoslovak climbers. *UIIA International Symposium.* Munich, Oct 1986.

Schussman LC and Lutz LJ. Mountaineering and rock climbing accidents. *The Physician and Sports Medicine* 1982; **10**(6) (June): 52–56; 58–61.

Stanek M, Rotman I, Skricka P, *et al.* Health complaints and finger deformities in Czechoslovak sport climbers. *Mountain Medicine Conference.* Prague, Oct 1988.

Tropet Y, Menes D, Balmat P, Pam R, and Vichare P. *J Hand Surg* 1990; **15A**(5).

Viel E, Gautier R, and Dusatois JL. Examen différentiel de la force préhensive du pouce, des doigts externes et de doigts internes. *Médecine du Sport* 1984; **58**(2): 86–90.

Training procedures and basic principles to avoid the risk of injury in modern free climbing

Amtmann E. Morphologische Grundlagen der Belastbarkeit von Knochengewebe. In: Cotta H, Krahl H, and Steinbruck K, eds. *Die Belastungstoleranz des Bewegungsapparates.* Heidelberger Orthopädie-Symposium 1980; **3**: 94–107.

Baumann W and Stucke H. Sportspezifische Belastungsen aus der Sicht der Biomechanik. In: Cotta H, Krahl H, and Steinbruck K, eds. *Die Belastungstoleranz des Bewegungsapparates.* Heidelberger Orthopädie-Symposium 1980; **3**: 56–64.

Berg A and Keul J. Kurz-und langsfristige Anpassungsvorgänge beim Krafttraining. In: Buhrle M, ed. *Grundlagen des Maximal- und Schnellkrafttrainings.* Schorndorf, 1985.

Bertschi D and Radlinger L. *Theoretische und empirische Untersuchungen zu sportartspezifischen Finger- und Unterarmverletzungen bei Sportkletterern.* Bern, 1986.

Burtscher M and Jenny E. *Most common complaints and injuries in free-climbers conditioned by training.* UIAA, Munich, 1986.

Clarke C. *Summary of rock climbing injuries.* UIAA Mountain Medicine Data Centre, London, 1984.

Franke K. Die Chondrophatie des Sportlers. In: Cotta H, Krahl H, and Steinbruck K, eds. *Die Belastungstoleranz des Bewegungsapparates.* Heidelberger Orthopädie-Symposium 1980; **3**: 254–258.

Krause R, Reif G, and Feldmeier C. *Overuse syndromes and injuries of the hand and the forearm in free climbing.* UIAA, Munich, 1986.

Radlinger L. Zur Problematik des Finger- und Unterarmtrainings der extremen Felskletterer. *Die Alpen* 1984; **12**: 517–519.

Radlinger L, Iser W, and Zittermann H. *Bergsporttraining – Kondition, Technik und Taktik aller Bergsportdisziplinen.* Munich, 1983.

Radlinger L. Konditionstraining für Bergsteiger. 1. Einleitender Teil: Aufwärmen und Dehnen. 2. Haupteil: Krafttraining. *Die Alpen* 1986; **2**: 62–65; **3**: 104–105.

Reif G. *Training of the flexor muscles of the finger by using shelves of different breadth.* UIAA, Munich, 1986.

Rotman I. *Overuse syndromes of the hand in top Czechoslovak climbers.* UIAA, Munich, 1986.

Specific aspects of cave rescue

Amphoux M, Bariod J, *et al.* Rapport d'expérimentation sur harnais spéléo. Châlain, rapport COMED FFS, 1984.

Faust P. Accidents et incidents en spéléologie [Thesis]. Nancy I, 1983.

Mallard M. Secours et prévention en spéléologie [Thesis]. Lille II, 1985.

Marbach G, Rocourt JL. Techniques de la spéléologie alpine. Chronaches, TSA ed. (F), 1980; pp. 253–254.

Moeschler O. Le spéléo-secours suisse. *Stalactite SSS* 1982; **32**(1): 5–10.

Probst R. Die Schwarzer Rettungsbahre. *Stalactite SSS* 1982; **32**(1): 16–20.

Ritter L. La médicalisation des secours en spéléologie [Thesis]. Toulouse, 1973.

Serra JB. Management of trauma in the wilderness environment. *Am Clin N Am* 1984; **2**(3): 635–647.

Vines T. Cave rescue management. *Emergency* 1981; **13**(6): 22–67.

Paragliding

Balet P and Fragnière A. *Le vol libre, notions théoriques.* Fédération Suisse de Vol Libre, Wetzikon, Switzerland, 1987.

Ledoux X. Le parapente peut être dangereux. *Vol Libre* 1988; **145**: 46–47.

Ledoux X. Traumatologie et secours d'urgence. Accidents de parapente. *Urgence* 1989; **8**: 16–221.

Reymond M, Gottrau P, Fournier PE, Arnold T, Jacomet H, and Rigo M. Traumatologie du parapente. *Communication presented to the seventy-fifth Congress of the Société Suisse de Chirugie* 1988, Montreux, Switzerland.

Index

Numbers in **bold** type refer to illustrations.

Abruzzi, Luigi Amedeo, Duke of the 13
Acclimatization
  acquired 43
  and altitude insomnia 51
  cerebral edema of 60
  loss of natural 65
  natural 43
  and nutrition 49
Acetazolamide
  acute mountain sickness 51–53
  high-altitude cerebral edema 61–62
  periorbital edema 57
Achilles tendon
  rupture 125–126, **139**, 147
  tendinitis 124–125, 127, **136, 137, 141**
Acidosis
  caused by acetazolamide 51
  in hypothermia 95
Aconcagua **50**
Acromioclavicular dislocations 144
Acute mountain sickness (AMS) 51–53
  cerebral thrombosis 58
  edema 57
  insomnia 50–51, 52
  portable hyperbaric bag 207, **263**
  psychopathology 64–65
  pulmonary thrombosis 58
  risk factors 54–56
  weight loss 49
Aerobic power 46–47, **47, 48,** 50, **54**, 55–56
Aerodynamics
  laws of 31–32
  paragliding 200–201, **253–255**
Air absolute humidity 45
Air density 45, **46**
Air rescue operations
  doctors 34–40
  helicopters *see* Helicopters
Alarm *see* Help, calling for; Radio
Alcohol 80
Alouette helicopters 17, 26
Altimeters 33
Altitude 43–66
  adaptation to *see* Acclimatization
  cerebral edema *see* High-altitude cerebral edema
  early scientific expeditions to 13–15
  hypoxia *see* Hypoxia
  insomnia 51
  world's highest inhabitants 14
Altocumuli, lenticular 32, **32**
Alveolar gas, sampling on Everest **7**, 15
Alveolar hyperventilation 52
American Medical Research Expedition to Everest 1981 (AMREE) **7**, 15, 44, 49–50
Amnesia 117, 172
Amputations 84, 86–87, 88–89
AMS *see* Acute mountain sickness
Anaerobic capacity 47, 50
Anglo-American Pike's Peak Expedition, Colorado (1911) **4**, 14
Ankle injuries 149–150, **178, 179**
Anorexia 51, 52
Apathetic–depressive syndrome 64
Apnea 50–51
Arachidonic acid cascade 79
Asphyxia 109
Aspirin 58
Ataxia
  acute mountain sickness 51, 52
  high-altitude cerebral edema 60–61
Aucanquilcha 14
Avalanches 100–112, **101–118**
  hypothermia case report 96–97
  personal experience of 99
  survival percentage of buried victims **15**, 22, 108
  victim detection devices 207, **261**

Barcroft, Sir Joseph **5**, 14
Barometric pressure 44, **45**
  measurement on Everest 15
Barotrauma 109
Base camp, optimal altitude 56
Bell 47-J helicopter **36**
Bergschrunds 112
Bernoulli's law 33
Bert, Paul 13
Bicycle ergometer **6**, 15
Biofeedback 89, **93**
Blisters 124, 126, **135**
  frostbite 86
Blood
  chemistry changes at altitude 14
  coagulation 57
  lactate 47, **49**
  viscosity 57
Body temperature, elevated 52
Bone scanning 76
Boots
  and frostbite 80
  skiing 136, **153–158**
  snowsurfing 161
  trekking **134**
Bradycardia 93
Brain damage 61, 65, 66
Bretylium tosylate 95
Broken legs 150–151, **180–183**
Buflomedil chlorhydrate 77
Bühler, Fritz 35
Burns from lightning strike 118, **129–131**

Calcaneal spur syndrome (plantar fasciitis) 125, **138**
Carbohydrates
  absorption at altitude 50
  high intake to alleviate symptoms of AMS 49
Cardiac arrhythmias 92–93, 94
Cardiopulmonary resuscitation (CPR) *see* Resuscitation
Cardiorespiratory arrest 112
Cave rescue 187–194, **233–245**
Cerebral damage 61, 65, 66
Cerebral edema
  high-altitude cerebral edema **59–62**, 60–62, 65
  in hypothermia 93
Cerebral thrombosis 58
Cerretelli, Paolo 15
Cerro de Pasco **5**, 14
Chamonix hospital 76–77
Champéry rescue team **35**
Chest harnesses 166–170, **200**
Cheyne Stokes respiration 50–51
Children
  broken legs 150, **181**
  frostbite 90, **94, 95**
  skiing 138–139
Cho Oyu 14
Chronic mountain sickness (CMS) 65
Clavicle 144
Climbing
  injuries from equipment 116, **127**
  sport 176–182, **214–225**
  training to avoid injuries 182–186, **226–232**
Clouds 32, **32**, 117
Colles' fracture 144, **169**
Communications 18
  see also Help, calling for; Radio
Compartment compression in frostbite 78
Compression chambers 62
Cornea 67–69, **69**
Cornices 105, **110**

Cough 58
Crevasses
hypothermia case report 98, **100**
rescues 112–116, **120, 122–126**
Cyanosis 58

De Quervain mortality curve **15**, 22, 108
Deep venous thrombosis 58
Dehydration during altitude climbing 45
Delusions 65
Dexamethasone
acute mountain sickness 52–53, **52**
high altitude cerebral edema 62
high-altitude retinopathy 63
Dextran 77, 91
Diazepam 51
Dill, D. Bruce 14
Direct face rescues **12–14**, 19, 21
Dislocations
acromioclavicular 144
frostbite 90
knee 148
hip 147, **173, 174**
scapulohumeral 143, **166, 167**
Distress, international signs **22, 23**, 26
Doctors
availability **38–39**
cave rescue 190, **237**
climbing, first aid kit 195–196, **246–249**
equipment 39, **41, 42**
mountain rescue 34–40
mountain rescue rucksack 197, **250, 251**
Dogs
avalanche 110, **116**
St Bernard 34
Douglas, C. G. **4**, 14
Downhill skiing 135–164, **151–193**
Drowsiness
acute mountain sickness 65
hypothermia 92–93
Drug abuse 80
Drugs
first aid rucksack for mountain rescue doctors 197
frostbite 91
GRIMM action pack 37

sedative 50–51
trekkers' first aid kit 198–199
which make skin sensitive to solar radiation 71
Dyspnea 52

Ears, frostbitten 79
Edema
cerebral 93
hands 57
high-altitude cerebral edema (HACE) **59–62**, 60–62, 65
high-altitude pulmonary edema (HAPE) 52, **57, 58**, 58–59
periorbital 52, **56**, 57
peripheral 52, 57
EHM (High-Mountain Military Academy) of Chamonix 35
Eiger North Face **12–14**, 19
Elbow injuries 146
Emotional excitation 64
Endocrine changes at altitude 49
Epidural hematoma **208, 209**
Epitympanic thermometer 111–112, **118**, 209, **267**
Ergometer, bicycle **6**, 15
Euphoric–impulsive syndrome 64
Evacuation 18–20, 21
Everest
American Medical Research Expedition 1981 (AMREE) **7**, 15, 44, 49–50
barometric pressure at summit 15, 44, **45**
first ascent without supplementary oxygen 14, 48
maximal aerobic power at summit 46
Operation Everest 15
physiological problems of climbing 14
Exercise, storm-bound climbers 58
Extracorporeal circulation 95, **99**, 112
Eye protection 68–69, **68, 71**, 71–72

Face, frostbite **72**
Falls onto the rope 165–170, **198–205**
Fasciotomy 87–89
Fat absorption at altitude 50
Feet
frostbite **73, 74, 77, 78**, 80
injuries 149–150, **178, 179**
sprains 125–126, 127
stress fractures 126, 127, **140**
trekking injuries 123–127
see also Boots
Femorofemoral extracorporeal circulation 95, **99**, 112
Femur 147, **175–176**
Fingers
frostbite **79–80, 85**
sport climbing injuries 176, 179–181, **216–224**
strength training 183–184, **228–231**
First aid
kit for climbers, mountain guides, climbing doctors 195–196, **246–249**
rucksack for mountain rescue doctors 197, **250, 251**
trekkers' kit 198–199
Flame hemorrhage 62, **63, 64**
Flash flooding 188
Fluid retention 52
Fluid intake 52–53, 58
Foehn 32, 33, **33, 34**
Forecasting
avalanche risk 104
weather 30–33, 205
Freeze–thaw–refreeze injuries 78, **81, 82, 84, 85, 91**
Frostbite 73–91, **72–93**
associated with lightning burns 119, **132**
classification 74, **73–80**
skiers 142
Frusemide 37, 53, 57, 197

Gauli glacier **21**, 26
Geiger, Hermann **9**, 17, 35, **37**
Glaciers 112–113, **119, 120**
aircraft landings **9**, 17, 26
Gauli **21**, 26
Kander **9**, 17
Gloves 80
Glucose loading 50
Glucose oxidation 49
Glycogen resynthesis 49
Grand Combin **18**
Grenoble, Hôpital-Sud 141
GRIMM (Medical Action Group in the Mountains) 36
action pack 37, **41**
courses in mountain medicine 40
Ground rescue operations 39, **40**
Groupe d'Intervention Médicale en Montagne *see* GRIMM

HACE *see* High-altitude cerebral edema
Haldane, J. S. **4**, 14
Hallucinations 65
Hands
edema 57
frostbite **75, 76**
skiing injuries 144–145, **169, 170**
see also Fingers
Harnesses, mountaineering 166–170, **200–205**
Head injuries 171–172, **206–209**
Heart rate 50–51, **51**
Heat stroke 70
Helicopters **8–11**, 17–20, 26–30, **24–27**
Alouette 17, 26

avalanche rescues 110
Bell 47-J **36**
evacuation 18–20
Lama 26
mountain rescue 34, **36**
night rescue 29
pioneers of modern mountain rescue **8, 9**, 17
reduced live load 28–29
rescue crew 39, **39**
transport 18
winch rescues **10–14**, 19–20, **26**, 28
Help, calling for
alarm **16**, 23–25
in an avalanche 106–107, **112, 113**
international signs of distress **22, 23**, 26
radio *see* Radio
Hematomas
epidural **208, 209**
intramuscular 132, **147**
Hemiparesis 60
Hemorrhage
flame 62, **63, 64**
retinal 55, 58, 62, **63–65**
ring **50**, 61
subarachnoid 61
subhyaloid 62, **65**
Henderson, Y. **4**
Heparin
cerebrovascular problems 58
frostbite 77
High altitude *see* Altitude
High-altitude cerebral asthenia (HACA) 66
High-altitude cerebral edema (HACE) **59–62**, 60–62, 65
High-altitude pulmonary edema (HAPE) 52, **57, 58**, 58–59
High-altitude retinopathy 62–63, **63–66**
High-Mountain Military Academy (EHM) of Chamonix 35
Hiking 123–134
Himalayan Scientific Mountaineering Expedition (1960–61) 15
Hip dislocations 147, **173, 174**
Horizontal rescue net **42**
Hormonal changes at altitude 50
Houston, C.S. 15
Humerus 144, 146, **171, 172**
Hyperbaric bag, portable 207, **263**
Hyperbaric oxygen chamber 89
Hyperventilation
alveolar 52
at altitude 50
hypothermia 93
Hypocapnia 93
Hypocapnia 45, 51
Hypothermia 92–98, **96–99**
associated with frostbite 78
avalanche victims 109, 112
falls into crevasses 98, **100**, 116
Hypoxia 13, 47
cause of acute mountain sickness 51
central nervous system 64–66
frostbite 78, **81, 82**
increased airway resistance 45
physiological reaction to **55,** 56
Hypoxic ventilatory drive 52, 54–56

Immobilization of injured person 192, **240**
Infarction, brainstem 61
Infection, secondary
blisters 124, **135**
fractures 190
frostbite wounds 77, 88
sunburn 70
Insertionitis 124–125
Insomnia 50–51, 52
Insulin 50, 94
International Civil Aviation Organization (ICAO) Standard Atmosphere values 44, **45**
Intestinal malabsorption 49–50
Intramuscular hematomas 132, **147**
Intubation
equipment 37, 197
in the field 21, 36, **38**
Isotope studies 87–88, **92**

Joint dislocations *see* Dislocations

Kander glacier **9**, 17
Karakoram Mountains 13
Kellas, Alexander 14
Ketanserine 77
Knee injuries 148, **177**
ligament injuries 152–155, **184–187**
sprains 152–154
Krebs cycle 47, 50
Kukuczka, Jerzy 49

Lactate 47, **49**
Lama (helicopter) 26
Lassitude 51, 52
Latitude and barometric pressure 44
Legs, broken 150–151, **180–183**
Leucocytosis 59
Lightning 117–119, **128–132**
Lower limb injuries 146–151, **173–183**

Martignoni, Fernand 17
Matterhorn 17
Maximal oxygen consumption measurement 15
Maximal voluntary ventilation (MVV) 45
Medical assistance *see* Doctors
Medical training
doctors 39
mountaineering professionals 40
paramedical 36, 40
Mental disturbances at high altitude 64–66
Mesencephalon 171, **206**
Messner, R. 14, 48, **50**
Metabolism at high altitude 46–47
Meteorology 30–33, 204
Microcirculation, impaired at altitude 57
Migraine 55
Moine, Jean 17
Monoskiing 160–163, **194–197**
Mont Blanc
ascent in 1787 **2**, 13
first helicopter landing on summit 17
Monte Rosa **3**, 13
Mosso, Angelo 13
Mount Everest *see* Everest
Mount Logan 15
Mount Makalu **6**, 15
Mountain rescue 17–40
Mountain sickness
acute *see* Acute mountain sickness
Chronic (CMS) 65
Mountain winds *see* Winds
Muscle
lesions 128–134, **142–150**
loss of mass at altitude 47, 49–50, 55
Musculotendinous ruptures 125–126, **139**
Myocardial contractility 15

Naftidrofuryl 77
Necrosis 79, **83, 84**
Neurological disturbances at high altitude 57, 64–66
Nicardipine 77
Nifedipine 59
Night flying 29
North face rescues 19
North wind 32, 33
Nose, frostbite 80
Nuclear magnetic resonance (NMR) spectrometry 76
Nutrition
at altitude 49–50
frostbite 90
Nystagmus 60

Obstacles, area of influence 32
Oedema *see* Edema

Operation Everest I (1946) 15
Operation Everest II (1985) 15
Ossifying myositis 133, **148–150**
Osteosynthesis 141, 151, **170, 182, 183**
Overhanging rock face rescues 19
Oxygen
  administration during sleep 51
  ascent without supplementary 14, 48
  chamber, hyperbaric 89
  maximal consumption measurement 15
  secretion 14
  transport cascade 43, **44**
  treatment of acute mountain sickness 52
  treatment of high-altitude pulmonary edema 59
Oxylog 208, **265**
Oxypuls 208, **264**

Pain control, frostbite 77, 89
Palpitation 52
Papilledema **59**, 60
Paragliding 20, 200–205, **252–260**
Paramedical training 36, 40
Performance at high altitude 46–47, 56
Perfusion equipment 37
Periodic breathing 50–51
Periorbital edema 52, **56**, 57
Peripheral edema 52, 57
Peritoneal dialysis 95
PGHM (Platoon of High-Mountain Gendarmes) 35, 116
Phenoxybenzamine hydrochloride 89, 91
Photophthalmia 67–69
Photosensitivity 71
Physiotherapy 88–89, 90
Piper Super Cub **9**, 17, 35, **37**
Pizzo, Dr Christopher **7**, 15
Plantar fasciitis 125, 127, **138**
Platoon of High-Mountain Gendarmes (PGHM) 35, 116
Pneumatic drills 114–115, **122**
Polycythemia 57–58
Probes 110, **117**
Protective clothing
  boots *see* Boots
  eyes 68–69, **68, 71**, 71–72
  skiing 138
  sunburn 71
Proteins 50
Psychotic symptoms 64–65
Pugh, L. G. C. E. 14–15
Pulmonary edema 52, **57, 58**, 58–59
Pulmonary embolism 57–58
Pulmonary hypertension, hypoxic 51
Pulmonary thrombosis 58

Radio
  radiotelephones **17**, 18, 23
  transceivers 23–24, 107, 110, **112**, 207, **261**
  transmitters **18**, 207
  see also Help, calling for
Red cell production 57
REGA (Swiss Air Rescue Guard) 35
Relay stations 18
Renin–aldosterone system 51
Rescue doctors *see* Doctors
Rescue nets 21, **42**
Rescue units **19, 20**, 25
  19th century 34, **35**
Respiration
  Cheyne Stokes 50–51
  water loss 45
Resuscitation 112
  in the field 36, **38**
  hypothermia 95, 96
Retinal hemorrhage 55, 58, 62, **63–65**
Retinopathy 62–63, **63–66**
Rewarming
  avalanche victims 112
  frostbite 76, 80–85, **86–91**
  hypothermia 94–95, **98, 99**
Rimayes 112
Ring hemorrhages **60**, 61
Rotor turbulence 31–32, **32**, 201–202, **255, 257**

Safety bindings
  skiing 138, **159, 160**
  snowsurfing 161, **196**
St Elmo's fire 117
SAMU (Urgent Medical Assistance Service) 35, 37
Saussure, Horace Benédict de **2**, 13
Scapulohumeral dislocation 143, **166, 167**
Schiller miniscope 209, **266**
Schneider, E. C. **4**
Scott, Doug 49
Searches in avalanches 107
Sedative drugs 50–51
Séracs 113, **120**
Shivering 92–93
Sit harnesses 166–170, **200–205**
Skier's thumb 155–159, **188–193**
Skiing
  avalanches 104, 106, **111**
  downhill 135–164
Skin
  cancer 70, 71
  grafts 88
  lightning strike 118, **128**
Skull fractures, depressed 172, **207**
Sleep at altitude 50–51, 52
Sleeping pills 50–51
Slow ascent 52

Smoking
  climbers 49
  frostbite patients 90
Snow blindness (photophthalmia) 67–69
Snow bridges 113, **121**
Snow crystals 100, **102–104**
Snow plumes **30, 31**, 31
Snowpack 102–103, **105–108**
Snowsurfing 160–164, **194–197**
Solar radiation 67–72
Solar reflection 67, **67**
Solar-powered radiotelephones **17**, 18, 23
Speleology 187–194, **233–245**
Spinal injuries 173–176, **210–213**
  horizontal rescue net **42**
Sport climbing injuries 176–182, **214–225**
Sprains
  foot 125–126, 127
  knee 152–154
Standard Atmosphere values 44, **45**
Storm-bound climbers 58
Stress fractures 126, 127, **140**
Stretchers, underground 191, **238, 239, 242–245**
Strokes 57
Subarachnoid hemorrhage 61
Subhyaloid hemorrhage 62, **65**
Sunburn 70–71, **70**
Sunglasses 68–69, **68, 71**, 71–72
Sunscreens 71
Survival tent, caves 189, **235, 236**
Swiss Air Rescue Guard (REGA) 35
Sympathectomy 89

Tachycardia 58
Tachypnea 52, 58
Temperature, elevated 52
Tendinitis 124–125, **136, 137**
Thawing liquid 116, **126**
Thermal parachute 207, **262**
Thermometer, epitympanic 111–112, **118**, 209, **267**
Thromboembolic episodes 57–58
Thrombosis
  cerebral 58
  deep venous 58
  dural venous sinus 61, **61, 62**
Thumb, skier's 155–159, **188–193**
Training
  to avoid climbing injuries 182–186, **226–232**
  of doctors in mountain medicine 39
  for high-altitude climbing 48–49
  medical for mountaineering professionals 40
  paramedical 36, 40

Transceivers 23–24, 107, 110, **112,** 207, **261**
Transient ischaemic attacks 57–58
Trekking 123–134
Tripod 115, **123, 124**
Turbulence **30, 31**, 31–33

Ultraviolet rays 67–72
Undulatory effect **32**, 32
Upper limb injuries 143–146, **166–172**
Urgent Medical Assistance Service (SAMU) 35, 37

Valley effect 32
Valley winds **28**, 31, 202, **258**
Vascular complications at altitude 57–58
Vasodilators 76–77
Ventilation during evacuation 21
Ventricular fibrillation 95
Venturi effect 33
Vertical face rescues 19
Visual impairment
- high-altitude retinopathy 63, **66**
- periorbital edema 57
- photophthalmia 67

Vomiting 51, 52

Warming *see* Rewarming
Weather conditions
- affecting helicopters **27**, 29, 30
- affecting paragliders 20, 202, **256–258**

Weather forecasting 30–33, 205
Weight gain
- acute mountain sickness 52
- peripheral edema 57
- as preparation for major expedition 48

Weight loss 49–50
Whiplash injuries 168, **205**
Whirlpool baths 86, 88–89, 90
Winch rescues **10–14**, 19–20, **26**, 28
Winds
- anabatic 32
- foehn 32, 33, **33, 34**
- mountain **29**, 31
- north 32, 33
- rotor turbulence 31–32, **32**, 201–202, **255, 257**
- turbulence **30, 31**, 31–33
- valley **28**, 31, 202, **258**

Xylose absorption at altitude 50

Zuntz equation 44, **45**
Zuntz, Nathan 13